THE
WEATHER
COMPANION

AN ALBUM
OF
METEOROLOGICAL
HISTORY,
SCIENCE,
LEGEND,
AND
FOLKLORE

THE
WEATHER
COMPANION

GARY LOCKHART

WILEY SCIENCE EDITIONS
John Wiley & Sons, Inc.
NEW YORK • CHICHESTER • BRISBANE • TORONTO • SINGAPORE

THE WILEY SCIENCE EDITIONS

The Search for Extraterrestrial Intelligence, by Thomas R. McDonough
Seven Ideas that Shook the Universe, by Bryon D. Anderson and Nathan Spielberg
The Naturalist's Year, by Scott Camazine
The Urban Naturalist, by Steven D. Garber
Space: The Next Twenty-Five Years, by Thomas R. McDonough
The Body In Time, by Kenneth Jon Rose
Clouds in a Glass of Beer, by Craig Bohren
The Complete Book of Holograms, by Joseph Kasper and Steven Feller
The Scientific Companion, by Cesare Emiliani
Starsailing, by Louis Friedman
Mirror Matter, by Robert Forward and Joel Davis
Gravity's Lens, by Nathan Cohen
The Beauty of Light, by Ben Bova
Cognizers: Neural Networks and Machines that Think, by Colin Johnson and
    Chappell Brown
Inventing Reality: Physics as Language, by Bruce Gregory
The Complete Book of Superconductors, by David Wheeler and Mark Parish
Planets Beyond: Discovering the Outer Solar System, by Mark Littmann
The Oceans: A Book of Questions and Answers, by Donald Groves
The Starry Room, by Fred Schaaf
The Weather Companion, by Gary Lockhart
To the Arctic: An Introduction to the Far Northern World, by Steven Young

PUBLISHER: Stephen Kippur
EDITOR: David Sobel
MANAGING EDITOR: Andrew Hoffer
DESIGN: Stanley S. Drate/Folio Graphics Co. Inc.

Library of Congress Cataloging-in-Publication Data

Lockhart, Gary.
   The weather companion : an album of meteorological history,
science, legend, and folklore / Gary Lockhart.
      p.   cm.—(Wiley science editions)
   Bibliography: p. 216.
   Includes index.
   ISBN 0-471-62079-3
   1. Meteorology—Miscellanea. 2. Weather—Miscellanea. I. Title.
II. Series.
QC870.L63   1988   88-6884
551.5—dc 19   CIP

Printed in the United States of America
   10 9 8 7 6 5 4 3 2 1

CONTENTS

FOREWORD / *vii*

PREFACE / *ix*

I. WEATHER PAST

Famous Weather Anecdotes / 3
Noah's Flood / 6

Ancient Weather / 11
The Tower of the Winds / 14

II. WEATHER TOOLS

Secrets of the Barometer / 19
Natural Barometers / 21
The Thermometer / 23

Natural Hydrometers / 27
The First Forecast / 30

III. WEATHER PHENOMENA

Weather in the Bathtub / 37
The Way of the Winds / 40
The Sound of the Weather / 43
The Smell of Rain / 47
Cloud Predictions / 50
Prophetic Skies / 53
The Rising of the Mountains / 57
The Ring Around the Moon / 60
The Cross in the Sky / 63
The Turn of the Tide / 66

The Moon and the Weather / 70
Faraday's Weather Updated / 74
Solar Weather / 77
Northern Lights, Southern
  Winds / 80
Earthquake Weather / 84
The Electrical Sky / 87
The Neon Sky / 91
Radio Weather / 93

IV. STORM WARNINGS

Thermals and Thunderstorms
/ 99
The Mighty Hurricane / 102
Strange Rains / 105
Lightning and Thunder / 108

Green Lightning Rods / 111
Wind, Water, and Weather / 114
The Sailor's Weather / 117
The Savage Blizzard / 120

V. WEATHER AND WILDLIFE

The Weather Fish / 127
Fishing Weather / 130
Hunting Weather / 132
Insects and Weather / 136

The Predictive Leech / 140
Bird Predictions / 143
Animal Forecasts / 146

VI. BOTANICAL WEATHER

Plants and Weather / 151
The Weather Plant / 155
The Oak and the Ash / 159

Practical Phenology / 162
Rain Trees, Rain Forests / 166

VII. THE WEATHER, YOU, AND ME

Human Weather Reactions / 173
Weather Dreams / 175
Arthritis Predictions / 178
Mystery of Weather Behavior /
180
The Winter Weather Game / 183
The Spring Weather Game / 186
Praying For Rain / 189
Ancient Rain Making / 194

Modern Rain Making / 196
The Great Floods / 200
Weather Words / 204
Indian Summer / 206
Making Money on the Weather
/ 210
The Weather Kite / 212

BIBLIOGRAPHY / 217

INDEX / 225

FOREWORD

The overcast skies on that particular day gave intense anticipation and thrill. Snow was in the forecast. Young Frank Bergmann's vigil was quiet, yet he observed in awe and with wonder the vastness of the sky above him. And, as the day turned into the stillness and cold of night, he continued to search the skies above for the first sign of the storm's arrival.

His world was lit by a lamppost on South Bend's Notre Dame Avenue. Fixation for hours. Continued quiet anticipation.

But now his reward, a long vigil appeased. One snowflake entered, dancing a ballet through the visible light. More began to enter the spotlight of his stage. And then, finally, a wild caper of flakes started falling through the illumination as if to herald the coming of the beauteous theme and its crowning glory of white and purity. It was quiet as he watched this seeming coronation of earth, until mesmerized into his own dark, quiet world of repose and dream. He was indeed content in knowing that the morning would greet him with the surprise he'd come to love in his own quiet secrecy.

Frank Bergmann's story truly represents the awakening of a new awareness among literally thousands of people throughout the United States who, until just recently, have not shared such simple tales of the love and appreciation of weather collectively with others. The development of the new Association of American Weather Observers has provided a much-needed forum for such communication. It opened a new door for him.

For, although they have lived with crystal clarity within the soul of the authors, such colorful visions had lacked an audience. Listeners were few and far between. Expression of one's special relationship with the elements was difficult at best. Others would be puzzled by such intensity. Few understood. Until now.

Amid a myriad of scientific strides in weather forecasting, data processing, and satellite meteorology, there is yet another fast-growing interest. For within each and every community, people from all walks of life are now seeking to enjoy and learn about weather and climate on a much different scale. Using non-technical language. On an amateur, face-to-face level. Indeed, on a natural level.

There is something very special about the person who truly sees and appreciates the atmosphere around him. The relationship strengthens the spirit. It promotes a better understanding of self and environment. And it softens the pain in our daily lives.

Author Gary Lockhart's approach to weather study provides a fascinating look at some special and historical relationships with our atmosphere, through stories and shared experiences. Written and graphically presented for all weather enthusiasts and naturalists to share, this book will enlighten you with the experiences of others. Experiences which will indeed enrich your appreciation of not only our wonderful weather world, but also how we view ourselves within it.

STEVEN D. STEINKE
Editor
American Weather Observer

(Steven D. Steinke and Frank Bergmann are both members of the Association of American Weather Observers, a not-for-profit organization devoted to the interests of amateur weather enthusiasts. Mr. Steinke is also the Editor of the organization's monthly publication, the *American Weather Observer*. For more information, write: AAWO, PO Box 455, Belvidere, IL 61008.)

PREFACE

The study of weather in relationship to human knowledge and belief is called ethnometeorology. In ages past and in many cultures it was part of daily experience to have some knowledge of weather lore, but today we depend on forecasts from radio and television. Weather prophets were common in the small towns of America, but they have long since disappeared and have been replaced by "weathermen" or meteorologists.

Near the Iowa farm of my youth was the Mississippi river town of McGregor. A few feet away from the river was a teepee where a Native American named Emma Big Bear lived. Each fall she was interviewed by the local newspapers for the winter report. Her predictions were probably based on cornhusks, the thickness of squirrel fur, and the time when the ducks flew south.

The rich traditions of Emma Big Bear's and our own ancestors have been replaced by satellites and computers. We do understand the weather better, and we have sophisticated, accurate forecasts. An understanding of natural forces and subsequent influences enriches our feeling for life and enables us to "tune into" nature. It is my wish that your skies and sunsets will never be the same.

Special thanks go to meteorologist Eric Wergin, whose friendly discussions lead to this book. Another word of thanks goes to Karen Murphy, who turned crude ideas into artful sketches. I have tried to keep the weather expressions of many poets and writers alive, through the use of old journals and records of antiquity. All translations have been put into modern English.

I.
WEATHER
PAST

FAMOUS WEATHER ANECDOTES

While living in the mountains, I saw that old farmers could predict both rain and sunshine, being right seven or eight times out of ten. I asked them how they did it, but they said it was only practical experience. If you ask people living in the cities, they don't understand this.

Since I had plenty of leisure time, I usually rose early in the morning, and then with an empty mind concentrated on the beauty of the fields, trees, rivers, mountains and clouds and I found that I could predict the weather right seven or eight times out of ten. Then I realized that in quietness the universe can be observed, the inner moods felt and real truth obtained.

YEH MENG-TE, A.D. 1156

There are several versions of an old story of Sir Isaac Newton's walk. "Beautiful morning," he said to a shepherd. "It will rain soon, sir," replied the shepherd. "I don't think so, the sky is almost clear," said Newton. "No," insisted the shepherd, "it's going to rain." Newton continued his walk, but an hour later he was soaked. Returning to the shepherd, he asked how he knew this. The man replied, "See that sheep over there? When she turns her tail to the wind it always rains."

Although Newton laid much of the foundation for modern science, he is not known to have taken an interest in the weather. It is said that Newton predicted there would be little difference between the summer and winter of 1682. He was right, for it was a cold, wet summer with a poor harvest.

Leonardo da Vinci took a more modern view of the weather. He noted that the winds were characteristic of the places of their origin. At a time when dew was believed to be of a supernatural origin, he believed that it was only water in the air. He noted that the ring around the moon was a sign of "vapors" in the sky.

Weather studies occupied the time of our first two presidents. George Washington began his notebooks in 1767 entitled, "An Account of the Weather." Thomas Jefferson kept copious notes on the weather, which were edited and published in 1944 under the title, *Thomas Jefferson's Garden Book.*

Benjamin Franklin did even more toward the study of American weather. He was the first to note that storms move from west to east, and that winds shift as the storm passes. In his library he had five thermometers and a barometer.

Franklin was the first to recognize that volcanoes affect the climate. During the severe winter of 1783–84, he correctly guessed that the cause of the cold was either a meteorite shower or the ash from the volcanic eruption at Hecla, Iceland. The reflection of sunlight by the ash lowered the winter tempertures.

Henry David Thoreau was the first American naturalist to be an avid collector of weather lore. The wind turning east meant rainy weather for him, and white fog over the pond was a sign of fair weather. The chirp of the robin spoke of rain, and the call of the loon meant wind. Thoreau mentions that plentiful acorns and thick cornhusks were signs of a cold winters.

One of the most interesting bits of weather lore concerns Christopher Columbus. He saw his first hurricane in August of 1494 and watched another one from the shores of Hispaniola in October 1495. By the time of his fourth voyage to the New World, he was familiar with native hurricane lore. He noted the large numbers of seals and dolphins on the surface of the ocean. Oily swell was coming from the southeast and veiled cirrus clouds were in the sky. There were light, shifting winds with a beautiful crimson sunset.

He sent Captain Terreros ashore to Governor Ovando of Santo Domingo, with the request that he be allowed to anchor his ship in the harbor, because of the coming storm. The governor mocked his request as the folly of a "prophet and soothsayer." Without permission to land, he took his ships to the mouth of the Rio Jaina, which gave him some protection from the wind. The rest of the fleet ignored his warnings and set sail for Spain. Nineteen ships were lost with all hands, and only one ship reached Spain without damage.

The enemies of Columbus declared he had raised the storm by magic. It disturbed Columbus, who wrote "What man ever born, not excepting Job, who would not have died of despair when in such weather, seeking safety for my son, brother, shipmates and myself, and we were forbidden the land and the harbors that I, by God's will and sweating blood, had won for Spain."

Perhaps the most interesting use of weather lore in modern times has been done by the Seminole tribe of Florida. They attracted much attention in 1926 when the entire tribe moved to another reservation north of the Everglades. They predicted that a seven-foot wall of water would sweep over the area, and during the storm, six feet of water did.

The night before the great storm, weather bureau forecasters were saying that the storm might strike Florida. The next day thousands of homes were destroyed and there was a heavy loss of life, but the Seminoles lost nothing. After the storm they heard that the white men were giving out free food, but when they saw how much damage they had suffered, they returned empty handed.

The tribe's first hurricane sign was an unseasonable blooming of saw grass. They knew that the height of the grass indicates the depth of the flood water. A week before the storm, rats and rabbits began moving north and westward. The birds stopped singing and began flying northwest. Alligators "barked" with unusual frequency and began moving into deep water.

The Seminoles continued to be accurate forecasters. In 1944, Florida was the target of two hurricanes. Before the first, the weather bureau issued an "all clear," but the Seminoles left the area. The storm turned around and struck Florida. For the second storm, the Navy and Coast Guard began evacuating planes and equipment, but the Seminoles didn't move. That storm missed Florida. Perhaps we should send our weathermen into the swamps of the Everglades to take lessons on hurricane forecasting from Nature herself. We need look no further than our natural environment to learn lessons about the weather.

NOAH'S FLOOD

The wind and flood raged for six days and nights, but on the seventh day, the stormy wind exhausted itself and died down and the flood water receded. I surveyed the scene and the earth was silent. Man and all his works turned to mud and clay. I opened a hatch and daylight fell upon my face. I wept as I looked for signs of life. On the twelfth day I could see a dozen patches of land sticking out of the water. The ship eventually grounded on Mount Nisir, and it stayed there for six days until we got out."

The Babylonian flood story. The Noah of the Gilgamish flood epic was Utanapishtim.

Nearly every year, newspapers publish accounts of expeditions going to Mt. Ararat in eastern Turkey to locate Noah's Ark. Films have been produced, and the ark has become part of public consciousness. Does Noah's Ark really exist on this 16,945 foot mountain?

In the biblical account, Noah was the only man on earth favored by God. At the age of five hundred, God gave him instructions for building a giant boat 3 stories high and 450 × 75 × 45 feet high. It took a hundred years to build the boat, and then pairs of all known birds and animals were taken on board. Heavy rains fell for forty days until the peaks of the highest mountains were submerged.

A half year later, the survivors landed on a mountain and waited until the waters receded to repopulate the earth. After sacrificing to thank God, they traveled east to the town of Babel and built a great tower to cele-

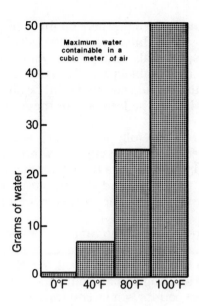

brate. God became angry and changed their common language into many languages.

Biblical dating is done by backdating the genealogies, and this has produced three dates for Noah's flood. Archbishop Ussher dated it at 2349 B.C., and these dates were added to the King James Bible in A.D. 1703. The Septuagint (Greek Old Testament) dated the flood at 2242 B.C., and the Rabbinical chronologies date it at 2103 B.C. This legendary flood must have happened between 2,100 and 2,350 years before the time of Christ.

In 1927, the British archaeologist Leonard Woolley excavated Ur of the Chaldees, the legendary home of the biblical patriarch Abraham. Digging forty feet below the surface, he found a layer of flood-washed debris ten feet thick. Further digging showed that this flood extended from modern day Baghdad to the Persian Gulf.

The entire area was within the flood plane of the Euphrates River, and most of it was less than twenty feet above sea level. The lower areas of Ur were only five feet above sea level, but the higher sections of the town were untouched. To those living in the flat lower valley, it must have looked as if the entire earth was flooded, when all they could see was flood water in the distance.

This flood was incorporated into a remarkable series of stories involving the Babylonian hero Gilgamesh. The inscriptions on the clay tablets parallel the Bible story, except for the fact that the forty days in the Bible are seven days on the tablets. Both stories end with a sacrifice and the gods smelling the odor and being pleased.

If a universal flood occurred around 2200 B.C., there ought to be plenty of geological evidence for it. There would be thick varves, great slack-water deposits, huge mounds of dead trees, animals, and plants. All of the old Egyptian documents and all Babylonian clay tablets would be ruined. Raised beaches would be carved into mountains. Pollen counts would be altered in swamps, and thick records would be left in the ice cores of the Arctic and the Antarctic.

Archaeologists digging through caves carefully study layers, which they call "horizons." These layers are caused by shifts in climate, winds, vegetation, and habitation. A universal flood would cause a thick layer of mud and organic materials to be deposited in the caves, and this would be readily recognizable and easy to date. Such an obvious layer has never been found, even at locations near sea level.

Numerous tree-ring records extend through the time of the flood, and no alteration of the ring sizes or of the isotopic ratios has been found. Ice cores from Greenland and Antarctica have been traced back 70,000 years. The layers show cycles and record the shifting climate, but they don't show a universal flood 4,000 years ago. Varve counts (silt layers) trace the history of many lakes back 10,000 years. These show many variations in rainfall patterns over the centuries. Pollen counts in marshes show exactly how the vegetation has changed in the past 10,000 years, and they trace the cold and warm periods as well as changes in plants. If Noah's flood was a world-wide event, evidence should be everywhere, but it can't be found anywhere.

The events after the flood are even more suspicious. Where did the three billion cubic miles of water go? How did Noah get the marsupials to Australia and New Zealand? How did genetics allow Noah to have a black son, a yellow son, and a white son? Supposedly only one language existed on the ark, but written records of the Sumerian, Akkadian, and Egyptian languages start before the era of the flood and continue for more than a thousand years afterward. They make no mention of a flood except as an ancient legend.

The early church fathers answered the questions on Noah's flood this way: "It's obvious, just look at the fossils in the mountains." Leonardo da Vinci examined the fossils in the mountains near Lombardy, Italy, around the year 1500. He observed that clams could only move twelve feet a day, far too little to reach the mountains. Under the title of "Doubt," Leonardo wrote in his notebook: "At this point natural causes fail us and therefore in order to resolve such a doubt we must either call in a miracle to our aid, or else say that all this water was evaporated by the heat of the sun."

The discovery of America raised a new series of questions. People wondered how the animals there got in the ark. Joseph Acosta wrote, "Who can imagine that in so long a voyage men would take the pains to carry foxes to Peru, especially the kind they call "acias", which is the filthiest I have seen. Who would say that they have carried tigers and lions? Truly it is a thing worthy of laughing at to think so."

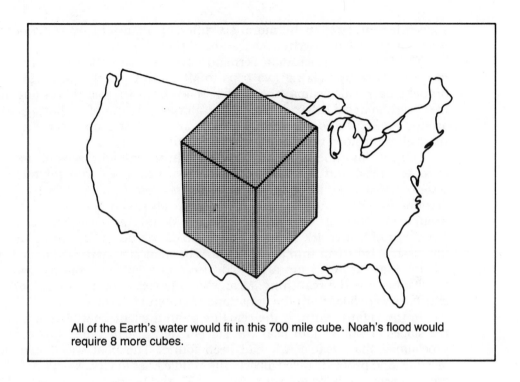

All of the Earth's water would fit in this 700 mile cube. Noah's flood would require 8 more cubes.

Sir Walter Raleigh spent the years from 1603–18 locked in the Tower of London writing *The Historie of the World* before he was executed. He raised the question of how Noah had room for all the animals. The Ark must have seemed huge to Bible writers, but they had never been to a small modern zoo.

The Ark is about the size of a modern 150-unit apartment building. If we assume that it was divided into 150 units, leaving space for the stairways, this leaves us 150 units measuring 25' × 25'. If we take two of each species, then each apartment has to have 115 birds, 69 reptiles, 57 mammals, and 5,000 insects. Each apartment would also have to have enough food and fresh water to last for a year.

Early world travelers found many flood stories, which added support to the biblical believers. But floods are experiences happening every year in some part of the world. Before the days of writing, our ancestors entertained themselves by telling stories. Local floods became world-class events when retold many times over the evening campfires.

The Bible account says that the survivors traveled westward and built the tower of Babel, which was excavated in 1899 by German archaeologists. Why are Ark seekers traveling four-hundred miles north of the Bible location to a mountain that was renamed "Ararat" by European travelers in the sixteenth century? Furthermore, six different mountains were named by Jewish and Christian writers as the site of the ark.

The French businessman Fernand Navarra began the modern search for Noah's Ark by making two trips to Mt. Ararat in 1955 and 1969. He brought back oak beams (the wrong wood) from the mountain and had them carbon dated by six different laboratories. Five out of six datings were around A.D. 650 and the sixth was A.D. 270. The timbers were cut 3000 years after the flood was supposed to have taken place.

During this time in history, a dozen churches had pieces of the true cross of Christ and many other relics. As a Christian, you might make several pilgrimages to ask for special favors from God. As proof of your devotion, you were expected to donate generously to centers with holy relics. Around A.D. 700 a group of clerics selected Mount Agri Dag (Ararat) as the "true" site of Noah's ark. Teams of men hauled oak beams up the mountain and assembled them into a large boat. The men may have originally been "rebuilding" the Ark from a few old pieces of wood. As time passed, the replica, became the real thing. Meanwhile the collection plates of the local churches were filled with the donations of grateful pilgrims.

In the eighth century, the remains of an ancient boat were found on the slopes of Mount Judi near the Tigris River. Learned men immediately proclaimed that Noah's Ark had been found. The boat was taken to a Mosque, and people traveled all over the Middle East to view Noah's Ark. The mosque was struck by lightning in A.D. 776 and burned, so that version of Noah's Ark is lost to history.

The Ark books present no up-to-date materials from geology, because there are none. Proof should come from the ice caps of Greenland and Antarctica, varves, pollen counts, horizons etc. Creation scientists have resorted to geological distortions. The Paluxy River layer that is supposed to show footprints of man and dinosaurs is seventy million years old, and has no relationship to strata existing 4,200 years ago. Some of the prints show unmistakable signs of being carved. Others are obvious solution marks and many have only a superficial resemblance to human prints.

The strength of the story of Noah depends on the question, "Where is the flood?" Those who travel to Mount Ararat in order to prove the words of Jesus, Mohammed, Moses, Peter, and Paul are mistaken. I think they must sleep uneasily at night remembering the old story of how the weavers wove the invisible cloth to make the finest garment for the emperor, only to have the grand occasion ruined by someone shouting; "The emperor has no clothes on."

ANCIENT WEATHER

600 B.C.: Thales attributes the Nile floods to wind changes.
500 B.C.: Phenological calendar assists farmers in China.
400 B.C.: Hipprocrates writes *Airs, Waters and Places*, a study of climate and medicine.
334 B.C.: Aristotle writes *The Meteorologica*.
300 B.C.: Theophrastus writes *On the Signs of Rain, Wind, Storms and Fair Weather*.
278 B.C.: Aratus writes the *Book of Signs*.
—Some Landmarks of Ancient Weather Knowledge.

Early man believed that the weather belonged to God. Storms were proof of God's anger and rain was proof of God's blessings. The earliest book of the biblical collection is the book of Job. After Job suffers and returns to God, he learns: "Great things doeth God, which we cannot comprehend. For he saith to the snow, fall thou on the earth; likewise to the shower of rain, and to the showers of His mighty rain. Out of the chamber of the south cometh the storm; and cold out of the north. By the breath of God, ice is given and the breadth of the waters is congealed."

The ancient Hebrews pictured a pie-shaped land mass indented by the Mediterranean and Red Seas. God stretched out a flat earth (Isaiah 44:24) within a great sea (Psalms 24:2). This sea had four corners, and many interpreters believed that since God had "bounded the seas," there was either a strip of land or a distant mountain range, so ships didn't have to worry about falling over the edge. The earth was supported on pillars, and when they shook (Job 9:6) there were earthquakes.

Today's Bible contains little weather information, for the book which gives a complete view of Hebrew meteorology was excluded. However, the book of Enoch was accepted by the early Christians and is extensively quoted in the book of Revelation. In it, Enoch visits heaven and learns the secrets of the calendar year, the moon's travel and the timing of the heavens. He also visits a prison where stars are punished for being out of time with the other stars.

Enoch finds that there are three doors on each side of the four quarters of the heavens above the earth. Angels guard giant storehouses of wind; when they open the doors the wind rushes out. The position of the doors determines the strength and direction of the wind.

11

There are storehouses filled with clouds, dew, frost, and hail released at God's direction. With such control, God is able to give good weather to his followers and bad weather to the unbelievers.

The ancient Greeks had a less ordered concept of heavenly weather, but they thought of the winds as being controlled by the gods. Mercury, the Messenger of the Gods, was also known as the Shepherd of the Clouds. The anger of Zeus resulted in the creation of storms.

While Hebrew writers created greater heavenly theology, the Greeks turned to philosophy in order to understand the natural world. The annual flood of the Nile aroused a great deal of thought. Thales of Miletus thought that strong winds held back the water during the dry season, and when the winds ceased in the fall, the floods came.

By 450 B.C. Empedocles developed the theory of the four combinations behind the universe. Water and fire were obvious opposites and so were air and earth. Their combinations were used to explain all natural phenomena and the Greeks tried to use these to explain the weather.

Democritus is known to chemistry students as the man who first explained matter as a combination of atoms. He used the analogy of the market place to explain the winds. When crowds meet in a narrow place they jostle one another for space and move slowly. In an open space the crowds speed up. In the same way winds move at speeds proportionate to their space.

Aristotle thought in terms of a pyramid of earth, water, air and fire. In reality, it gets colder with increasing altitude, but Aristotle believed in a layer of fire high above the earth. This fire explained the northern lights and heat waves.

Theophrastus became the successor to Aristotle, and continued to teach Greek philosophy. He developed more rational explanations for weather phenomena. He found that winds coming from the sea were more likely to be associated with rain, and clouds passing over mountains would drop the water they contained. After his death, meteorology fell into a 2,000 year silence, which was broken when Torricelli invented the barometer.

The first known map of the earth is a clay tablet now in the British Museum. The "Mappa Mundi" shows Babylon in the center of a round land mass centered in a square ocean. A similar map was created in bronze around 500 B.C. by the Greek philosophers Anaximander and Hecataeus. The Greek historian Herodotus wrote "They draw the world as round, as if fashioned by compasses, encircled by the river of ocean." He doesn't disagree with this view and tells the story of the Libyan sailors who took three years to sail around Africa. They had the "sun on their right hand," a proof that they were on the other side of the equator and that the world was round. Herodotus says that he simply can't believe this.

If the book of Enoch had been included in the biblical canon, we might have had "creation meteorology" with a flat earth and heavenly storerooms filled with clouds and hail. However it may have been so rare, that when the Council of Nicea put together the scriptures, no council member present had a copy.

The early church fathers were advocates of the flat earth. Some, such as Saint Augustine, presented both theories and admitted that it was impossible to tell which was correct. Others such as Chrysostom commented on the Book of Hebrews 8:1 and added: "Where are those who say that the heaven is in motion? Where are those who think it spherical? For both these opinions are here swept away."

The third book of Lactantius, "On the False Wisdom of the Philosophers" was written around A.D. 320. He ridiculed the absurdity of the belief that there were places where the rain and snow fell upwards and people have their heads above their feet. Around the year A.D. 540 Kosmos wrote a book called, *Against Those Who, While Wishing to Profess Christianity Think and Imagine Like The Pagans that the Heaven is Spherical.* He quoted extensively from the church fathers and stated that "no man can serve two masters," when it came to believing the accepted religious truths about the earth.

The fall of the Roman Empire in A.D. 476 produced a more liberal group of theologians willing to consider other ideas. Pope Sylvester II, who ascended the throne in A.D. 999, was the first church leader to accept a spherical earth. The church still retained its dogmas and Galileo stood before the Inquisition because of his belief that the earth rotates. To his theory they offered an irrefutable answer: "Joshua commanded the sun to stand still, not the earth."

Christopher Columbus was not completely certain that the earth was round, but he was inspired by Ezra IV, 6:32 which stated that 6/7ths of the earth's surface was covered by water. He reasoned that the earth was bounded, so he had little to worry about as he set sail in search of a new passage to India.

Flat earth theology was accepted by some groups of fundamentalist Christian believers until as late as the mid-twentieth century. South Africa was dominated by the Dutch Christians, the Boers, who continued to hold a literal view of the Bible. It was after a brief visit to South Africa during the Boer war, that Rudyard Kipling wrote his short story, "The Village That Voted the Earth was Flat."

The first person to sail around the world alone was a tough old Maine sea captain called Joshua Slocum. In 1898 he sailed his sloop "Spray" to South Africa and stopped at Port Natal for a brief rest. He was introduced to the president of South Africa, Paul Kreuger, now immortalized on one of South Africa's monetary units, the Kreugerrand. When he told the president that he was sailing around the world, Kreuger interrupted, "You don't mean around the world! It is impossible! You mean in the world. Impossible! Impossible!"

THE TOWER OF THE WINDS

But soon a tempestuous wind arose called the Euraquilo and came from the land, and the ship was caught and could not face the wind, and we gave way and were driven. After running under the near shore of a small island called Cauda, we managed with difficulty to secure the boat; and after hoisting it up, they took measures to undergird the ship. Fearing that they should run on Syritis [a shoal to the west of Cyrene] they lowered the sails, and so were driven onward. As we were storm-tossed, they started throwing cargo overboard; and the third day they threw overboard the tackle of the ship . . . St. Paul said: "Men you should have listened to me and not set sail from Crete."

ACTS 27:14

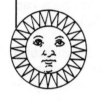

Is there a secret place where weathermen meet to worship the God of the Skies? That secret place is known to archaeologists as the "Horologion of Andronikos." It is only a fifteen-minute walk from downtown Athens, and just a short distance from the far more famous Parthenon.

Also known as the "Tower of the Winds" it is made of white Pentelic marble and stands forty-six feet high. If it were taller and situated on a prominent hill, it would be considered one of the world's great monuments. The Parthenon and many other famous Greek antiquities have suffered terribly from the ravages of the weather and man. But the Tower of the Winds remains almost as it was built fifty-years before the time of Christ.

The Babylonians were the first people to have an eight-sided wind rose. This symbol was adopted by the Greeks and incorporated into their architecture. Each of the eight sides of this building has a sculpture that personifies the characteristic of each of the eight winds. Here are their directions and descriptions:

North: Boreas is a warmly clad man with a conch shell.

Northeast: Kaikias is a man carrying a vessel filled with small round objects that may represent hail.

East: Apeliotes is a young man carrying an armload of fruit and grain.

Southeast: Euros is an old man wrapped securely against a rainy wind.

South: Notos carries a water jar with the neck downward.

Southwest: Lips is a lightly clothed boy gripping the ornament of a ship.

West: Zephyros is a youth with a flower pot.

Northwest: Skiron is a bearded man holding a firepot.

Tower of the Winds

Andronikos, the architect of the tower, should receive the title of the world's first scientific weatherman, for he was the first to link time and wind direction. The sides of the tower were inscribed with lines like a sundial to indicate the time of day. Inside the tower was a water clock so the time could be read on cloudy days. The roof of the tower contained the first known weather vane. However, it may not have been accurate because of the deflection of the winds from the Parthenon and taller buildings.

Julius Caesar is believed to have appropriated funds for this remarkable building around 50 B.C. It is mentioned by the Roman writer Varro in 37 B.C. Vitruvius, another Roman writer, wrote about the weathervane on the roof. It was a bronze Trident holding a wand pointed in the direction of the prevailing winds.

In Roman times, the building stood in the marketplace of Athens. The library of Hadrian and the Roman Agora were nearby. It is quite possible that sailors and merchants met in the building before setting sail because of its clocks.

Did the Greeks think of the winds as gods, or were they merely winds? We know from Homer's Odyssey (1200 B.C.) that changes in the winds were due to the will of the gods. Homer wrote,

> He [Poseidon] gathered the clouds and troubled the water of the deep, grasping his trident in his hands; and he roused all storms and all manner of winds and shrouded in clouds the land and sea and down came the night from heaven. The East wind and the South winds clashed, and the stormy West and the North, that is born in the bright air, rolling onward a great wave. . . .

By the time of Aristotle (330 B.C.) the wind was attributed to the repulsion of the sun. He used the reasoning that the sun repels rain and snow under its path. The strongest winds gather at the middle of storms, and their clashing creates storms.

Oddly enough, the principal Greek god of the winds is not depicted on the tower. It was the legendary harp of Aeolus that produced the sound of the breeze, and his blowing on the conch shell that produced gales. According to Strabo, the Greek historian, Aeolus was once an ordinary seaman who was honored because he taught sailors how to navigate the dreaded strait of Sicily where the whirlpool of Charybdis made shipping dangerous.

The historian Diodorus wrote that Aeolus was a king who invented sails and established storm signals. The blind Greek poet Homer has an interesting account of Ulysses meeting him. "Then we came to the Aeolian Island, and there dwelt Aeolus Hippotades dear to the deathless gods; there he dwelt in a floating island and round it was a wall of brass that could not be broken; and a smooth rock cliff. Aeolus gave the winds to Ulysses bound in a leather bag with a cord of silver."

The Tower of the Winds is one of the world's greatest mystical representations of the elements of the weather. The early Christians once used this building to meet; later, whirling dervishes held their ceremonies here. If man ever starts a "weather religion," this is surely the place for its headquarters.

II.

WEATHER TOOLS

SECRETS OF THE BAROMETER

Captain to cabin boy: "How's the barometer?"
Cabin boy: "Rising sir, steadily rising."
Captain: "And my brandy?"
Cabin boy: "Falling sir, steadily falling."

(A SAILOR'S JOKE)

And what else? In the glass the barometers tell us,
Just how soon the winds will blow.
And how soon the heavy rain will beat down on the grainfields,
Or, when the clouds are driven away, the sun will dry them.
Our hopes will no longer be deceived in this,
The glass lends a hand and the thing is done.
It tells us exactly how the heavenly bodies move.
Through it a force difference, is revealed in the weather.

—MICHAEL LOMONOSOV, A POET AND FOUNDER OF THE
RUSSIAN ACADEMY OF SCIENCES, WAS OVERLY ENTHUSIASTIC
ABOUT THE BAROMETER AROUND 1752.

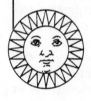

Aristotle was the first person to suspect that air had weight. He filled a leather bag with air and weighed it, then he pressed the bag flat and reweighted it. Since he did not detect any difference, he concluded that air had no weight.

The old Greek philosophers believed that "nature abhors a vacuum." They noticed that when a tube of water was closed at one end and tipped upright, the water refused to run out, but if a hole was punctured, the water quickly ran out. The philosophers did not understand that we live in an ocean of air that is the equivalent of thirty-four feet of water covering the entire surface of the earth. We are like fish who cannot realize the nature of water until they are removed from it.

Galileo should have invented the barometer, for he experimented with long tubes of water. When the tube was above thirty-four feet, the water would "break like a rope." The significance was lost to Galileo, but his student Torricelli got the credit for inventing the first barometer in 1644.

The final confirmation that the barometer was really measuring the atmosphere came on September 19, 1648, when the French scientists Blaise Pascal and Florin Perier got together. Perier climbed the Puy-de-Dome with a new, mercury barometer and it fell three inches in the 4,800 foot climb. Pascal was delighted to receive the first positive proof that the barometer was measuring atmospheric weight.

19

During this time Ferdinand II, the Grand Duke of Tuscany, organized the first group to study the weather. He expected to find high barometric pressures on rainy days, for after all, the sky was full of water. He was surprised to find that the barometer dropped before storms, and the rate of the drop was related to the speed of the winds. When scientists began to understand atomic weights, the mystery of the light weight of the air was explained. Water (H_2O) in vapor form has an atomic weight of 18, but a molecule of oxygen (O_2) has an atomic weight of 32.

The first known barometric prediction of a storm was done by Otto Von Guericke in 1652. He used a glass tube to construct a 34-foot barometer. Floating on the top of the water was a "weather mannikin" which pointed to the lines on the tube. Although he gained a good deal of fame from this prediction, his barometer broke before the storm arrived.

In 1664 Robert Hooke observed a barometer over several years. He noted that the barometer sank in relationship to the severity of the storms, but in 1 out of 10 storms the barometer actually rose. At this time the instrument maker George Sinclair graduated a barometer beginning at 28 inches with *tempest, stormy, much rain, rain, changeable, fair,* and *long fair.* These markings have been traditional ever since on most English barometers.

The greatest problem faced by the early barometer makers was that not all towns were at sea level. In Denver, the barometer would rest at a *storm* setting in beautiful weather. Eventually, a standard sea-level barometer was developed, and all barometers were calibrated for altitude, so that each location was equivalent.

The weatherman looks at the equal pressure lines (isobars), and the way they make patterns of good weather (highs) and (lows). Meteorologists also determine the isobars of barometric pressures at higher elevations to gain further clues to the patterns of future weather changes. Since we now live in well-lighted homes, few of us realize that barometers glow in the dark when shaken. It was noted that some, but not all of the early barometers would glow in the dark. In 1723, Charles Du Fay found that in order for a barometer to glow, the mercury had to be absolutely dry.

One of the most interesting observations of the barometer was made by an old sea captain. He did not believe in watching the fall of the barometer when bad weather threatened. He watched the top curve of the mercury known as the meniscus. If the meniscus were rounded like a marble, good weather was on the way. If it were flat, a storm was coming. Inevitably he was right!

Water clings to glass more strongly than it adheres to itself, so its meniscus is curved upward. Mercury has a greater surface tension, so it curves downward in a glass tube. There may be unknown factors influencing the top of the mercury, but most of us will be content to forecast the storms by the height of the mercury.

NATURAL BAROMETERS

The signs which the sun and the moon give are necessary to human life, and those who learn to read the signs, can benefit like the wise, who have a long experience in using observations. The weather can be indicated by many signs which people can use to know about rain, drought and winds. The extension of the disk of the moon and the circle around the sun means heavy rains or strong winds without doubt. If by the third day the moon is clear and shining, it is a sign of good weather, but if her crest is sharp and red, it indicates heavy rain and strong winds. Who can ignore all these observations which are so useful for living life? The navigator who watches the signs does not worry about returning safely to port. The traveler who takes note of changes in the sky takes care to avoid the effects of bad weather. The farmers sowing grain and cultivating the crops can choose the most favorable time for their work.

—SAINT BASIL

Long before the first mercury barometer was constructed, nature was indicating barometric changes. Since we are measuring pressure, anything enclosing air will expand when the pressure falls.

Air rushing out of caves becomes quite noticeable, when the barometer drops. The "Wind Cave" of the Black Hills and the "Medicine Hole" of the Killdeer Mountains of North Dakota are two places that were believed to have special significance in early times. I have seen outcoming air shaking the bushes near vents along the Mississippi River. The caves inside the cliffs must have been of considerable size to expel such large quantities of air.

It is not unusual for deep wells to serve as barometers. Near Megrin, Switzerland, a whistle was put on the opening to a deep well. When there was no sound, the weather was stationary. When the whistle blew, the weather grew worse. An inblowing whistle sound indicated good weather.

Another way of using a well to indicate weather changes is to put a sack on the casing. The bag will bulge before bad weather, and will be sucked inward in good weather.

Some coal mines are known as "singing mines." There are trapped gasses in the rocks, which are released as the mine is extended. When there is a low pressure area coming from a high wind, which creates a lowered pressure over the mine, the miners hear a faint sound like buzzing bees or bagpipes.

Generally speaking, the difference between high and low pressure areas is less than 3%. Since our atmosphere is an equivalent of 34 feet of water, the suction effect is equivalent to 1 foot of water. The low pressure area

21

exerts the same force against the rock walls as it takes to suck water up a 12-inch straw. With miles of tunnels and poor ventilation, the extra gasses can make a big difference in the mine.

A century ago English authorities investigated the problem of mine explosions. They believed that up to half the explosions were due to low pressure. Mine authorities were required to have barometers, and shut down mining operations if pressure dropped to dangerous levels. Better ventilation cured the problem.

It was found that the Kimberly mines of South Africa, where diamonds were mined, had a high concentration of methane gas. A vent was drilled along the mine area to release the gas. During days of good weather and high pressure, there would be a flame several inches long. During barometric drops or stormy weather, the flame would grow by several feet.

The formation of scum on ponds and lakes is a common indication of lowered barometric pressure. There is a physical phenomenon known as the "cartesian diver." A slight pressure change will send an air bubble hurrying to the top or bottom. As the barometer falls, these bubbles of gas carry organic matter with them and form scum. Ponds with lots of marsh gas will have a layer of black decaying matter on their surface before bad weather.

It has often been noticed that springs flow more before a storm and water becomes discolored in wells. The pressure drop of a severe storm could raise water 2 to 3 feet higher than normal.

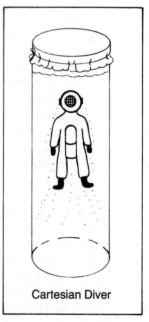

Cartesian Diver

Most of the changes that occur in the heights of water in wells are not a result of the barometer, because many wells respond to the moon. Japanese scientists found that wells in Tokyo rose and fell twice a day, in the same rhythm as the tides. The water level in the wells changes from 1 to 3 centimeters depending on the height of the ocean tides.

Perhaps the most interesting natural barometers are the geysers of Yellowstone National Park. A geyser is a column of superheated water. Many of them erupt at regular intervals, but others are irregular. They depend on subterranean heat, the greater the heat flow, the shorter the period of eruption.

The "Splendid Geyser" of the Daisy group is weather sensitive. It normally does not erupt unless the pressure drops before a storm, and then it shoots hot water 100 to 190 feet into the sky. It continues to erupt while the barometer remains low. The lower the barometer, the shorter its eruption period. When the barometer rises, it becomes quiet again.

Nature is mainly barotrophic, that is, it responds to the barometer's drop. Mosquitos bite harder and fish bite less. We feel slightly more sleepy, and we may dream more. If we have arthritis, we may feel like living barometers.

THE THERMOMETER

Each day I have worked on and changed the instrument for measuring temperature and have done this, so that if I could be with you and talk with you, I would tell you about the history of the invention and of its improvements. Since you first wrote to me, and I believe that you have invented it, and I believe that your instruments are constructed with great artistry that far surpasses mine.

—A LETTER OF SANTORIO SANTORII TO GALILEO

There was no thermometer before the year 1600. Galen, the great Roman physician, used a temperature scale of 4 degrees each of cold, normal, and hot. Usually people described temperature in terms of cold, chilly, comfortable, warm and hot.

The ancient Greeks almost invented the thermometer when they noticed that a hot liquid rises more than a cold liquid in a glass tube. The great difficulty in constructing hollow tubes was probably the reason they didn't invent the thermometer.

Around the year 1600, four people stumbled onto the thermometer. It seems likely that Galileo made the first air thermometer around the time he became a professor at Paduo, Italy, in 1592. This seems to have been a tube of glass bent into a circle with the temperature expressed at the coldest degree of 120 degrees representing (a mixture of salt and snow) and a temperature of 360 degrees expressing (the hottest days in summertime). Twenty years later, his student Sagredo mentions it to Galileo as "the instrument for measuring heat which you invented, but which I have made in several convenient styles."

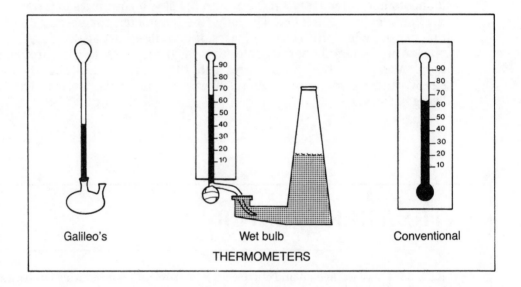

Galileo's   Wet bulb   Conventional

THERMOMETERS

The original barometers were sealed at the top and open at the bottom. The thermometers were open at the top and closed at the bottom. The problem with these early instruments was the scale. It took almost a century to realize that water always freezes and boils at the same temperature, and this could become a standard for marking temperature scales.

When Galileo died in 1642, Prince Leopold gathered his students into a school known as the "Academia del Cimento." The pupils used "Florentine" thermometers, which were sealed at the top. The thermometer liquid was made of "spirits of wine" dyed with kermes. The scale of these thermometers went from roughly 20° in winter to 80° in summer. These were used throughout Europe for the next hundred years.

The two milestones in thermometers were the first use of mercury by Athanasius Kircher in 1620, and the first division of the thermometer scale in 1679 by Sebastiano Bartold using the freezing and boiling points of water. These points vary with altitude, so they are accurate within limits.

Nature has several interesting thermometers. The speed at which ants crawl is in direct relationship to the temperature. In 1897 Professor A. E. Dolbear wrote an article entitled, "The Cricket as a Thermometer." "Dolbear's Law" was $T = 50 + (N - 40)/4$. "T" is temperature and "N" is chirps per minute. Another way of determining cricket temperature is to count the number of chirps in 14 seconds and add 40. Since the cricket is probably shaded by grass, his temperature may not be the same as yours.

If you have no thermometer you might consider using a rattlesnake. The frequency of rattling falls to zero just above the freezing point and rises to 100 cycles at 98°F. The rattling frequency rises by 1½ rattles per second per degree Fahrenheit. If your rattlesnake is rattling at 60 rattles per minute, subtract ⅔ from that number. Add the 40 to 32°F (freezing) and you have a temperature of 72°F.

The angle of rhododendron leaves makes a fairly good thermometer. At normal summer temperatures they are horizontal, but as the temperature approaches freezing they are about half way down, and at zero, they hang downward. The leaves began to curl at 35°F, and at 20°F the curl is about the diameter of a thumb and at zero it is about the diameter of a pencil, with individual plant and species variations. The Rhododendron maximum is the most cold-hardy, and will survive winters in the North Central United States.

The Euonymus radicans coloratus is planted for the beauty of its foliage. The leaf color is said to be a natural temperature indicator. As the temperature drops it changes from a light red to a deep red and then to black.

The thermometer is not a predictive instrument like the barometer, because temperature varies throughout the day. If you wet the bulb of a thermometer, the temperature drops as the water evaporates. A wet bulb thermometer has a lower reading than a dry bulb, except when it rains and the air is saturated with water. The reading between a dry and a wet bulb thermometer gives the "relative humidity," the reading that is predictive of future weather.

There is quite a bit of weather lore that chimney smoke going upwards predicts good weather. The Scientific Society of Rochdale, England, kept a record of smoke observations. When smoke rose vertically, there were 42 fine days with 6 wet days. If the smoke hung around the chimneys, the chances of wet to dry days were 21 to 6.

The rise of chimney smoke is related to the rate of air temperature drop. In wet weather, the temperature varies little; but during good weather, the temperature falls as the height increases. This rate of temperature drop is known as the "lapse rate." Hot smoke rises faster and further when the temperature drops off with altitude.

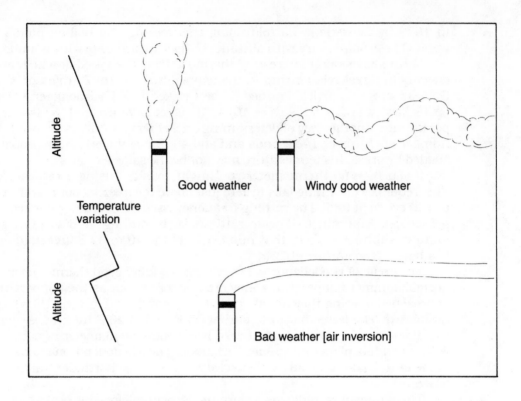

Dry, warm chimney smoke loses 1.6°F per 300 feet of altitude and wet saturated smoke loses .8°F per 300 feet of altitude. The smoke continues to rise as long as it is warmer than the surrounding air. Smoke loses heat by mixing and by expanding as it rises. It will rise until it becomes equal to the surrounding layer. In bad weather smoke rises until it reaches the "inversion layer." The layer is warmer than the smoke, and since the smoke can't rise through it, it spreads out into a layer.

NATURAL HYDROMETERS

When the clouds are on the hills, they'll come down by the mills. When Chevriot ye see put on his cap, of rain ye'll have a wee bit drap. If Roving Pike do wear a hood, Be sure the day will ne'er be good. When Bredon Hill puts on his hat, ye men of the Vail beware of that.

—THIS OLD ENGLISH POEM REFERS TO THE FACT THAT AS THE RELATIVE HUMIDITY RISES, CLOUDS FORM OVER HILLS AND MOUNTAINS.

Hydrometry (measuring water) is the science of determining how much water air can hold at a given temperature and pressure. This determination shows us the probability of clouds, fog, and rain.

There is no "absolute humidity," there is only relative humidity and it is relative to air pressure and temperature.

Long before scientists learned the trick of measuring relative humidity, several types of hydrometers were used. On the Chiloe Islands, off the coast of Chile, the natives used the shell of Lithodes antarcticus. This crab is related to our king crab, and when dry the shell is a light gray color. The shell is about 3 by 4 inches, and was hung on the walls of the native huts. When the amount of moisture in the air increased, the shell becomes spotted with red splotches. As the humidity increases, these enlarge until the entire shell is a dark red color.

The Chiloe Islands are dry during the summer, which lasts from October to May. During the months of April through September, the winds come from the north and the rainy season begins. It is during this time that the crab shells turn red.

Both the Australian aborigines and English farmers have made use of seaweed to determine the probability of rain. Many types of kelp contain large amounts of magnesium chloride. This chemical is hydroscopic, that is, it absorbs water vapor from the air. When the relative humidity rises, the kelp will feel damp long before the air feels damp.

There are other chemicals which change color when damp. Some snails are yellow before rain, and blueish afterwards. One of the best known rain predictors is cobalt chloride. When paper is soaked in this chemical, it turns blue when the relative humidity is less than 40%. The color changes to lavender at 40 to 55% and pink above 55%. This property is the basis of a weather toy sold under the name of "chameleon barometer." Paper flowers soaked with cobalt chloride are also sold as humidity indicators.

One of the most noticed signs of high relative humidity is sweat on glasses. In his *Natural History* Pliny the Elder wrote, "Whenever you see at

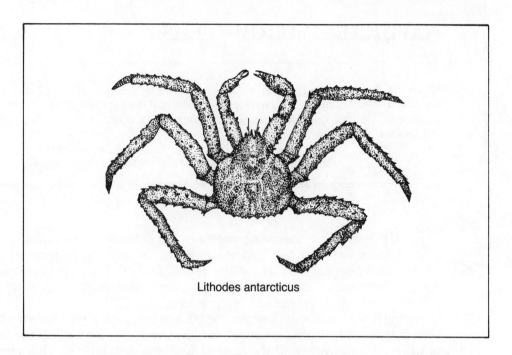

Lithodes antarcticus

any feast the dishes and plates on which food is served, covered with dew, be assured that it is a token of terrible storms approaching."

The old sailors noticed that an increase in relative humidity caused their ropes to shrink, and this tightened up the masts. The sailors referred to the rays of light spilling through evening clouds as "the sun tightening up his backstays."

The engineers who moved the Egyptian obelisk known as Cleopatra's Needle to St. Peter's Cathedral in Rome were said to have miscalculated the distance that the rope would have to be tightened in order to raise the great rock needle. The pillar was almost up, but the winches couldn't be tightened any more. The workmen were ordered to wet the ropes, and this shortened them enough to lift the obelisk into place.

The same phenomenon happens with our hair. When it gets damp, hair shortens by about 2%. This isn't much, but it is enough to make hair into a sensitive and accurate device for recording relative humidity.

The first attempts to make hydrometers from human hair were inaccurate, because of the natural oils on hair. The French geologist William DeSaussure boiled hair in a sodium carbonate solution for a half hour and then used the hair for his instruments. It is much simplier to wash the hairs in ether, and then use them in hydrometers.

DeSaussure placed his hairs in "extreme dryness" and then in "extreme moisture." This enabled him to calibrate his hydrometers to measure from 0 to 100% relative humidity. He had a great deal of conflict with fellow

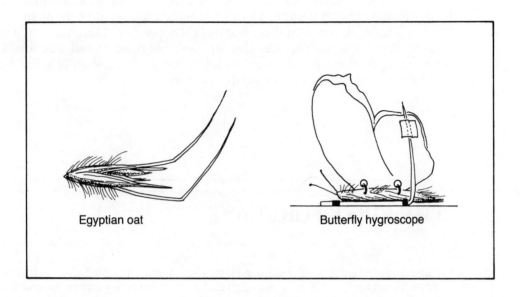

Egyptian oat

Butterfly hygroscope

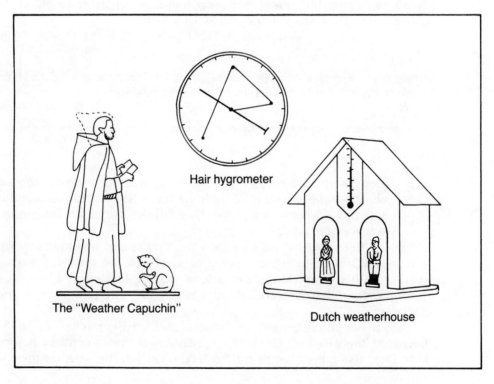

Hair hygrometer

The "Weather Capuchin"

Dutch weatherhouse

scientists, who didn't believe that this was accurate. It was accurate, for hair has a fairly linear rate of shrink as the relative humidity rises.

Hair hydrometers are the basis of a number of weather toys. Hair controlls the "Dutch Weather House," in which a man appears at the door in bad weather and a woman appears in good weather. The shortening hair inside the weather monk pulls the cowl over his head in bad weather. The shortening of our hair in damp weather may make it slightly more curly before rain and straighter during dry weather.

THE FIRST FORECAST

Probably northeast to southwest winds, varying to the northward and westward and eastward and points between. High and low barometer swapping around from place to place, probable areas of rain, snow, hail and drought, proceeded or preceded by earthquakes with thunder and lightning.

("THE WEATHER") THIS IS MARK TWAIN'S IDEA OF A NORMAL
NEW ENGLAND WEATHER FORECAST.

Whenever two Englishmen meet, they began to talk about the weather, each trying to inform the other of that which they know nothing about.

—DR. SAMUEL JOHNSON

Men have been forecasting weather for millions of years from natural observations. It is a little more than a hundred years since the first scientific forecast was made. How did the transition from nature to science take place?

The oldest surviving regular daily weather record was made by William Merl of England. During the years of 1337–44, he recorded each day's weather in his diary. His observations weren't like ours. There were no thermometers, barometers, wind gauges or rain gauges for another 300 years.

When the Royal Society of London met on September 7, 1663 they discussed the weather. Dr. Wilkins mentioned that a century before, Dr. John Dee, also a member of the Society, observed the weather for 7 years, and acquired such skill in predicting it, that he was known as a witch. It was suggested that a history of the weather should be recorded, and Robert Hooke was appointed to this position. He fulfilled this job admirably by inventing an instrument to measure wind speeds and he studied barometer readings in relationship to the weather.

The forecasters lacked one essential ingredient: a way of gathering distant observations. This was fulfilled by the invention of the telegraph, and in 1848 it was suggested that a telegraph station be put on the most distant part of Ireland. Since weather systems flow eastward, an indication of the weather could be obtained.

In 1849, the Secretary of the Smithsonian Institute asked telegraph operators to begin each morning by transmitting a single word on the weather such as "clear," or "rain," and so on. These reports were gathered in Washington, D.C., and then grouped into a simple weather chart, without any attempt to forecast weather.

The first storm predictions were issued in the Netherlands in June 1860. The Dutch meteorologist Christopher Buys-Ballot used his laws of the winds and the barometer to find the centers of low pressure systems. By linking barometric observations from several towns, he was able to find the direction and speed of storms.

Many of us are familiar with Charles Darwin's *Voyage of the Beagle* and his observations that lead to *The Origin of Species.* The captain of the *Beagle*, who so favorably impressed Darwin, was Robert Fitzroy. When Commander Stokes died in 1828, he was promoted to captain. Due to his skill he made the long and dangerous voyage with Charles Darwin during the years of 1831–36.

As a reward for his careful work, Fitzroy was appointed governor of New Zealand from 1843–45. His defense of the rights of the native Maoris against the English settlers lead to such an outcry that he was recalled to England. He was promoted to Admiral and given the job of developing a weather warning service in 1859. He began by dividing England and Ireland into 3 districts with telegraph stations. These observations were sent to "Lloyds of London," where they were on display for ships' captains. "Lloyds" was a coffee house where ship captains and shipping men met before going to sea.

In August of 1861, Fitzroy felt so confident of his skill, that he began sending reports to the newspapers. He called these reports "forecasts," and wrote that they were not prophesies, but opinions established from science. The word "forecast" may have been so little used, that he thought he was the first to apply it to the weather. This is not quite true, for the first known use in English occurred in 1533. Le Berners wrote "A shipmaster forecasteth and is in great thought and feare of tempests and stormest to come."

Fitzroy developed 47 rules from observations to further the prediction of the weather. In 1862, he put all of his studies into *The Weather Book.*

Robert Fitzroy

During the following years, he saw Germany, Holland, and Russia following his example by establishing forecasting services.

His daily forecasts kicked up a storm of criticism at home. The Royal Society felt that he was making predictions without enough scientific data to do so. When he died in 1865, the Royal Society stopped all forecasts for a few years. There was a popular outcry from sailors and farmers who depended on the forecasts.

There appears to be a dark side to Fitzroy's feelings that we cannot really know. The young naturalist whom he had taken with him on the Beagle thirty years before, published his acclaimed, *Origin of Species.* Fitzroy was a deeply religious man, who is said to have preached sermons in fundamentalist churches telling how he sought to warn Charles Darwin of his mistaken ways. He sent out letters under the pen name of "Senex" criticizing Darwin's book. Darwin's defenders pulverized his arguments. In April 1865, the father of weather forecasting cut his throat with a shaving razor.

Admiral Fitzroy did not make any major contribution to weather knowledge. He almost discovered the theory of air masses in 1863 by noting that depressions formed at the meeting of air streams with different properties. He did bring together all of the scattered observations into a useful weather service.

Fitzroy's predictions were limited because of the technology of his time. This limitation was overcome when Marconi sent his first radio signals across the English Channel in 1899. The United States Weather Bureau immediately began a program to make practical radio receivers and transmitters. By 1904, the U.S. Navy was transmitting severe weather warnings to ships. Nine years later they were broadcasting a daily weather bulletin. The first daily public weather forecast was made by the University of Wisconsin's radio station on January 1, 1921, and the era of Fitzroy's dream came true.

Weather forecasting is the one aspect of science that nearly everyone uses each day. We do our fair share of complaining about this difficult art. If we ever choose to honor it by having a "Weather Day," I would propose Fitzroy's birthday, which is July 5.

III.
WEATHER PHENOMENA

WEATHER IN THE BATHTUB

The wind continues in the old point Southwest, which independently of detaining us appears invariably to bring bad weather with it. The ship is full of grumblers and growlers, and with seasickness staring me in the face as bad as the worst. The time, however, passes away very pleasantly, but instead of working, the whole day is lost between arranging all my nick-nackeries and reading a little of Basil Hall's fragments."

CHARLES DARWIN'S DIARY FOR DECEMBER 15, 1831.
THIS IS THE BEGINNING OF THE *VOYAGE OF THE BEAGLE.*

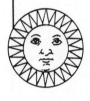

Suppose you were kidnapped and placed on an airplane, and sent to an unknown location. Days pass, and you are blindfolded and ushered into an underground cell. "Where am I?," you ask, but no one will answer your question.

There is a way that might give you a clue to find out your location. Fill the bathtub with water and pull the plug. As the water drains, it will swirl counterclockwise in the Northern Hemisphere and clockwise in the Southern hemisphere. If you are near the equator, it will not develop a swirl readily.

The forces that make water swirl are so minute that the geometry of the bathtub and the pulling of the plug can easily start the opposite swirl. The first person to do the bathtub experiment to demonstrate the rotation of the earth was the Austrian physicist Ottokar Turmlirz in 1908. The water had to sit for several hours to eliminate any "filling" motion. He used a 5-foot circle of water colored with dye to demonstrate that the moving earth will produce a force in water.

Scientists called the rotational force "Coriolis" after Gustave Coriolis who wrote a scientific paper in 1835 on the relationship of wind to pressure. He theorized that the swirls of storms received their direction because of the earth's rotation.

The first person to provide actual proof of the rotation of the earth was Joseph Furtenbach, a mathematician from Ulm, Germany. Galileo had recently announced that the earth rotates in space, discovered through his telescopic observations of Jupiter's moons. In 1627, Furtenbach fired a cannonball vertically into the sky, and then sat on the muzzle of the cannon. Since the earth rotates the ball should drop west of the cannon.

We do not see it, but Coriolis is everywhere. If a gun is fired at a target 400 feet away, the bullet will drift about a tenth of an inch to the right. If you drive down the road at 60 mph, the car drifts 15 feet to the right per mile. You never know this because of the friction of the tires. Since an airplane is not bound by friction it must correct for the Coriolis force.

The largest corrections are made by the military. During World War I, the German gun known as "Big Bertha" bombarded Paris from 70 miles away. The shells took 3 minutes to reach Paris and drifted a full mile to the right. When a missile is fired, the corrections necessary to put it on the target may amount to several hundred miles.

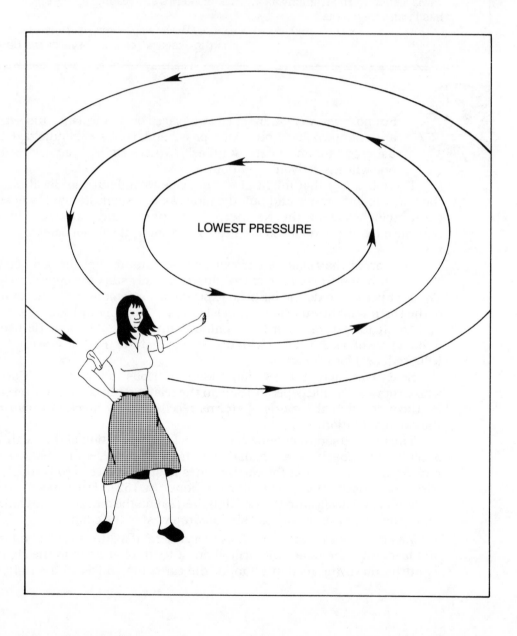

LOWEST PRESSURE

The best illustration of the Coriolis force may be had by playing catch on a merry-go-round. The ball has to be corrected for speed and direction or the other person will completely miss it. It can then be seen that we are not dealing with a "force," but a relationship. The ball is moving straight, but we are changing in relationship to its path.

What does Coriolis have to do with the weather? One example is found in the ocean currents. The drift moves the surface water away from the California coast in a southwest direction. Cold water then rises from the ocean depths and produces the notorious California fogs.

The Greeks observed the winds changing directions during storms, and they thought that stronger winds were influencing the storm. Modern meteorologists refer to storms as collisions of air masses. The interactions of air masses and the earth's rotation produces the swirl of the storm.

The Chinese were the first to recognize the wind change but they didn't get it right. In 1694, they wrote in the "Annals of Taiwan," "If it [a typhoon] blows from the north, then it shifts to the east, if from the east then to the south, and so on until it has gone through a complete circle. The "Tai" does not stop until the wind has gone around the entire circle."

The first person to study storm winds was the Connecticut engineer William Redfield. He traveled through Massachusetts after the great storm of 1821, and noted that the trees were blown over in a different direction from the trees near his home. By collecting further information he found that the storm had a counterclockwise swirl. Redfield predicted that storms in the southern hemisphere would circle clockwise.

Buys-Ballot used the direction of the swirl to find the center of the storm. In 1858, he formulated a law that winds are perpendicular to the lines of barometric slope. In 1860, he put his observations into a paper entitled, "Some Rules for Predicting Weather Changes in the Netherlands." In that paper was his well known rule, "with your back to the wind, low pressure is to the left." In the southern hemisphere, you face the wind to find the low pressure center with your left hand. The application of this law has enabled many sailors to head for calmer waters.

Gustav Coriolis didn't realize the implications of his famous law, or the fun that physicists would have in confirming it. Recently U.S. Navy scientists confirmed this effect at the South Pole by using a drum of ethylene glycol to keep from freezing. (A humerous application of the law is that guests at cocktail parties tend to circle clockwise around the buffet table.) It's a high-pressure area! But one thing is certain; Coriolis revolutionized our understanding of the weather.

THE WAY OF THE WINDS

I investigated everything including the road of thunder and lightning. The angels showed me the keys and the guardians of the storehouses and the way they operate. They carefully open them with a chain, so they don't wreck the angry clouds and destroy everything on earth.

I saw the treasure houses filled with snow, cold and frost; and I observed the angel with the seasonal keys. He fills the clouds with these, and the treasure houses still remain full.

I saw the resting places of the winds and observed the key bearers with scales and measures. They weight the winds and allow them to escape in measured amounts, lest the great force make the earth rock.

—*THE SECRETS OF ENOCH*

Neither the Greeks nor the Hebrews thought of the wind in a scientific way. The winds were supernatural, and connected to the gods or angels. Since the weather was controlled by God, strong winds were His punishment. When the Greek poet Homer spoke of a storm, he wrote, "All the winds conspired together."

The earliest of the Homeric epics speak of four winds. By the height of Greek civilization this had increased to eight directional winds. By the time Charlemagne ruled France, the four winds had acquired the cardinal names North, South, East, and West. Around 800, Charlemagne combined the cardinal names with directions to produce hybrids such as NNE for north by northeast. In order to give greater directional accuracy, the degrees of compass headings were adopted as directional readings in 1949. Mariners now speak of going west as a 270 degree heading.

In ancient times early man had little sense of direction, other than natural landmarks, and the sunrise and sunset. These points vary 23° to the north in midsummer and 23° to the south in midwinter. The first directional points were based on the position of the sun. The Italian word for *east* is *levante*, which means sunrise, and west is *ponente* or sunset. All English words for the cardinal directions come from the old Germanic languages. East (*austra*) means sunrise, west (*vis*) is sunset or rest. North (*vairan*) means dampness and south (*sunno*) is the place of the sun. In the Slavic languages, north means midnight, and south means midday.

The first sailors to use the compass supported it on a reed or reed cross, in a bowl of water. By the year 1302, a decorated card with markings was attached to the backside of the compass. This "wind rose" quickly became a rose, and the "fleur-de-lis" became the standard way of marking the north point.

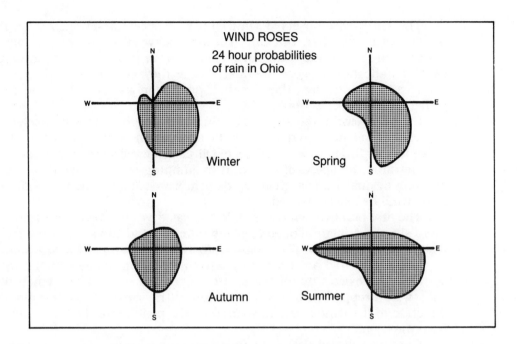

Meteorologists adopted the "wind rose," but they attached a different idea to it. It became the way of showing seasonal weather and the prevailing wind directions. Wind rose diagrams are done with rain and snow to show local weather patterns.

Here is a wind rose plot done for the state of Ohio at two different times of the year.

Chance of rain or snow in 24 hours.

Wind direction	N	NE	E	SE	S	SW	W	NW
January	50%	77%	42%	62%	61%	44%	45%	38%
July	9%	14%	47%	55%	60%	50%	40%	12%

The Ohio weatherman who put 10 years of study into these plots of wind and moisture, also observed a steady weather pattern over a given 24 hour period during autumn. He noticed that when the autumn's night wind came from the south there was a 94% chance of rain within 24 hours. Similarly, if a south wind came in the morning, rain followed only 10% of the time. And he observed that a southeast wind at night is usually followed by a rainy morning, but a southeast wind in the morning is not a sign of rain. An east wind at night during the summer is followed by rain by 88% of the time during the next day.

The old sailors observed wind shifts in order to set their sails and predict future weather. *Backing* (counterclockwise) winds were a sign of low pressure areas and a nearby storm. *Veering* (clockwise) winds were a general indication of good weather. Since they were constantly alert to changes in sail position, their weather prediction was done by wind shifts rather than wind direction.

Sir Isaac Newton remarked that, during his boyhood, his first scientific experiment was done to determine the speed of the wind. The great storm of September 3, 1658, was the day of Oliver Cromwell's death. Newton tried to determine the speed of the wind by jumping with the wind and then jumping against it. The great windstorm was a "foot stronger" than any other wind he had measured.

The first practical wind speed device was developed by Robert Hooke. It was a simple flat plate allowed to swing into the wind and scaled for velocity. A summary of all the wind-measuring instruments would fill a large book.

The Aztecs had an interesting wind measuring device which may be seen at the Mexico City museum. It consists of a tripod from which little balls were dropped into a series of concentric rings. The strength and direction of the winds were measured by the position of the balls dropped from the tripod.

Once we believed that the winds were due to angel-controlled gates at the four corners of earth. Now we understand that the rotating earth produces swirls in its airy cover. During World War I, Jacob and Wilhelm

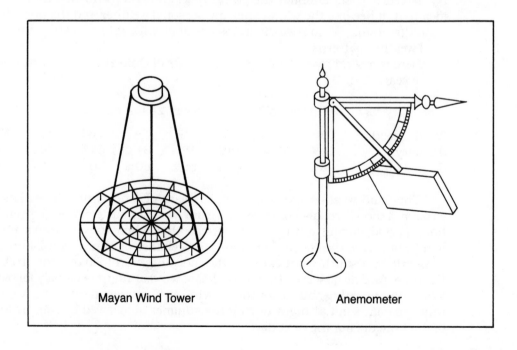

Mayan Wind Tower   Anemometer

Bjerkness developed the idea of air masses. The winds could be thought of as "continents" of air, and the storms were a result of "fronts" of the air masses. The forces that drive the air masses, producing the winds, are of great interest to weather forecasters.

THE SOUND OF THE WEATHER

The sounds which the ocean makes must be very significent and interesting to those who live near it. . . . I was startled by a sudden loud sound from the sea, as if a large steamer were letting off steam by the shore, so that I caught my breath and felt my blood run cold for an instant, and I turned about, expecting to see one of the Atlantic steamers thus far out of her course, but there was nothing unusual to be seen. . . . The old man said that this was what they called the "rut," a peculiar roar of the sea before the wind changes, which, however, he could not account for. He thought that he could tell about the weather from the sounds which the sea made.

Old Josselyn, who came to New England in 1638, has it among his weather signs, that "the resounding of the sea from the shore, the murmuring of the winds in the woods without apparent wind, showeth wind to follow."

Being on another part of the coast one night since this, I heard the roar of the surf a mile distant, and the inhabitants said it was a sign that the wind would work round east, and we should have rainy weather. The ocean was heaped up somewhere at the eastward, and this roar was occasioned by its effort to preserve its equilibrium, the wave reaching the shore before the wind. . . .

HENRY DAVID THOREAU, "CAPE COD"

When we hear the rumble of jet aircraft overhead, we are reminded of the change in the weather. The old proverb states, "Sound traveling far and wide, a stormy day will betide." The nature of sound and the weather is more complex than the proverb.

A set of church bells 5 miles from Lebekke, Belgium, are known as "water bells." If you can hear them clearly, rain is supposed to follow. This is partly true, for there is an optimum humidity for the air to carry sounds without dissipating.

The great English clock known as "Big Ben" is also known as a weather bell. This 14-ton bell was cast in 1854 and its 400-pound striker rings the hours for miles around. But hearing Big Ben is not so much a result of the weather as the wind. People living 3 to 5 miles northward take its sound as a sign of rain. Whereas people who live the same distance southward hear the bell and expect good weather. They are both correct. As a low pressure area crosses London, winds flow from the south and carry the bell sound further northward than it would normally be heard. The normal wind currents of a high pressure area carry the sound further southward, and thus accounts for the weather predictions.

The mysteries of ship wrecks in dense fogs are also of great interest. It has been found, in some cases, that ships were wrecked in normal range of the fog horns. Sound will often bounce off fog patches and produce an echo. Patchy fog destroys sound, but a fog of uniform density transmits sound well. The British Navy made a series of tests with a standard fog horn under various weather conditions. They found that when the humidity dropped from 77% to 71% the distance the fog horn could be heard dropped by almost 2 miles. This increase in the transmission of sound before storms explains the old proverbs.

There is another type of sound that indicates rain. The Roman poet Lucian wrote, "Nor less I fear from that hoarse, hollow roar, in leafy groves and on the sounding shore." The rain sound is usually heard near mountains or cliffs. Acoustical engineers call this roaring noise "vortice shedding." When air blows over a rounded object it swirls one way and then the opposite. The roaring noise depends on the speed of the wind, and generally foretells a storm in 6 to 12 hours. As the proverb says, "When the forest murmurs and the mountain roars, then close your windows and shut your doors."

In the forest we hear the whisper of the pines in the winds. The Japanese call this "the song of the pines" or "matsukaze." The tones produced by the wind are a result of the average diameter of the pine needles and the branches. It is the principle of the aeolian harp, and the tone depends on the speed of the wind.

There is a popular Estonian song that speaks of the storm noise and the belief that Finnish people are good at controlling the weather.

> Wind of the cross! Rushing and mighty.
> Heavy the blow of thy wings sweeping past!
> Wild wailing wind of misfortune and sorrow,
> Wizards of Finland ride by on the blast.

The old belief that telegraph wires can predict the weather has been subjected to several scientific studies. There is a theory that the wires pick up the storm vibrations hours before the storm arrives on the scene.

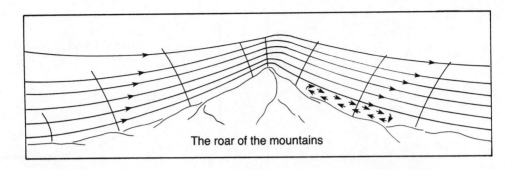

The roar of the mountains

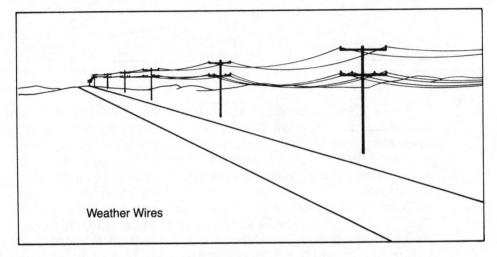

Weather Wires

Experiments show that the sound is due to the wind alone, but a gust of wind can activate the wires for several minutes.

Quite a number of people have noted the change in the sound of the wires and tried to delineate their meaning. Although the tones have to be heard, here is a summary of "wire predictions."

> Light humming: damp weather
> Deep humming: light rains
> High, shrill notes: brief heavy rains
> Buzzing tones: weather changes
> Increased EW hum: temperature fall
> Increased NS hum: temperature rise

In arctic regions it is a well-known fact that sound travels great distances. Animals in these areas have small ears, for they can hear a barking dog up to 15 miles away. In hot desert areas where sound travels poorly, animals have large, sensitive ears to maximize sound reception.

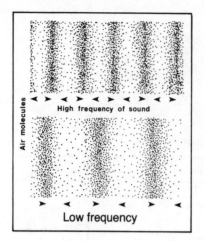

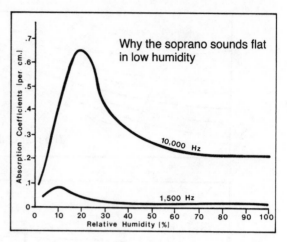

I once lived in a log cabin at an altitude of 11,000 feet in the Colorado mountains. At the mouth of the canyon, 3,000 feet downward and 10 miles away, was a rock quarry. On the cold, windless fall mornings I could hear every word the workers were saying. But by the time the sound reached me, it was merely puffs of noise, the phonetic discrimination was lost.

Occasionally you will see a musical review in which the reviewer describes the poor quality of the sound. This may not be the musicians' fault. The University of California at Los Angeles did a study of the acoustical properties of a concert hall. When the hall was filled with dry desert air at 15% relative humidity a 4,000 hertz (Hz), sound lasted 2.5 seconds. When humid ocean air filled the hall, the same sound lasted 4.5 seconds.

The scientists also found that a high-frequency sound of 10,000 Hz was absorbed 7 times faster than a 1,500 Hz sound in dry air. The worst possible setting for a soprano singer is a hot, dry concert hall. Ideal concert weather would be cool and humid. When you can clearly hear the sound of faraway trains and jet planes, that's the best time to go to a concert.

THE SMELL OF RAIN

The shepherds say that the sweet smell that accompanies the rainbow is noticeable especially when there are briars and brambles and shrubs which have a sweet scent. The reason for the sweet scent is the same as the earth smell, for whatever grows out of it, is sweet scented to begin with. For all things containing moisture have a sweet scent when they are warmed, for the sun absorbs the moisture.

—ARISTOTLE, "PROBLEM 12"

The day is hot and muggy and the wind is gusty. Suddenly you look up and say "It smells like rain!" Across the prairie comes a line squall with boiling clouds and streaks of lightning.

We do not smell rain, for water has no odor, although we do note the increase in dampness before the coming shower. However, there is a distinct smell, which used to be attributed to the rainbow. Both Aristotle and Pliny mention the ancient belief that rainbows have a smell.

The first speculations about this odor were made in England a century ago. It was believed that soil molds liberated odor when the barometer dropped. French scientists confirmed that there was an earth odor when they extracted 7 pounds of soil with solvents, and obtained an earthy smelling substance.

Australian chemists began to study the earth odor during the 1960s. They found that certain types of clay exuded a strong rain smell when the relative humidity reached 80%. They named the smell "petrichor" from the Greek words for "stone-essence." They set out to test the idea that the rain smell was due to micro-organisms in the soil. Chemicals from the clay had no optical activity indicating that they were not produced by soil bacteria.

In many parts of the world there is a distinct blue haze in the sky. The Great Smoky Mountains and the Blue Ridge Mountains of Tennessee are named after this blueish haze. The forests of eucalyptus trees in Australia grow in the Blue Mountains. It is estimated that plants exude 450 million tons of blueish haze into the atmosphere. These hazes contain a wide variety of complex chemicals from plants and trees.

The Australian scientists tried trapping the oils in the atmosphere. When concentrated and sniffed, they have a terrible smell. Then they added the oils to clay soil and allowed them to remain for several weeks. When the soil was distilled, the earth odor was clearly detectable, and the nature of petrichor became obvious. During periods of drought, the soil traps oils from the air and converts these oils into the familiar "rain odor." It appears that iron in the soil is the catalyst that converts the plant oils into the rain

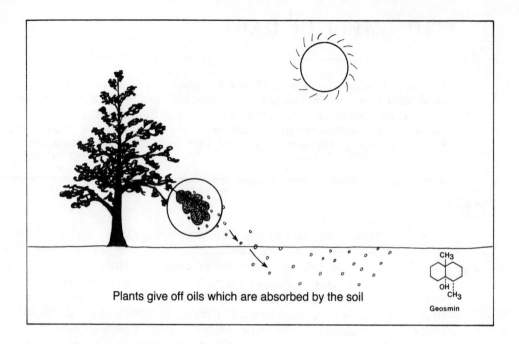

Plants give off oils which are absorbed by the soil

Geosmin

odor. If it rains regularly the rain smell will not develop because oils are washed away and there is no catalytic activity. When the relative humidity goes above 80%, the moisture begins to liberate the earth rain smell.

The scientists believed that petrichor might be a mysterious growth promoter, because plants and mushrooms spring up overnight from the rain parched soil. When petrichor was added to young plants, the substance retarded their growth, but there could be growth promoters among the mix of chemical substances.

There is a small perfume industry based on the rain odor in Kannauj India west of Lucknow. Workers put small clay disks outdoors during the months of May and June. These absorb air oils and then the smell is steam distilled from the disks. It is sold under the name of "matti ka attar," meaning "earth perfume."

The scientists found that a foot-square tray of clay picked up only about 3 milligrams of oils from the air. An analysis of a complex mixture of oils turned out many simple fatty acids. There were also esters such as methyl octaneate and methyl noneate. The chemists do not seem to have finished their analysis, so the various types of "dry land" rain odors cannot be described in their chemical variations.

The rain odor in wet country is of a different chemical composition. It is produced by organisms in the soil such as Streptomyces griseus and Streptomyces odorifer. These fungal organisms are similar to the ones that produce the antibiotic streptomycin.

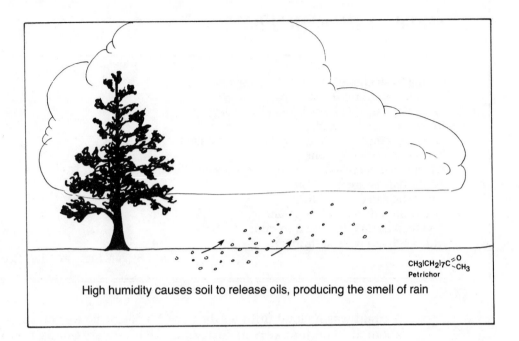

$CH_3[CH_2]_7C \overset{\nearrow O}{\underset{\searrow CH_3}{}}$
Petrichor

High humidity causes soil to release oils, producing the smell of rain

This moist earth odor is called "geosmin" from the words "earth-smell," and it shows up in several common smells and tastes. When heavy rains flood water supplies, the "off" taste in water is often produced by geosmin. In streams that have the algae Symploca muscorum, trout develop a muddy flavor, which disappears if they are kept in fresh water before being processed. Geosmin is one of the components of the muddy taste of catfish and carp, which feed on the bottom. The earthy smell of freshly cooked beets is mainly due to geosmin.

Advanced techniques in chemical analysis allow scientists to determine the chemical structures of the fungal earth odors. These are a mixture of four related chemicals, each having a distinct smell. These can be described as earthy, cinnamonlike, and camphoric. Together they produce the familiar earth odor that you smell when the barometer drops and rain is coming. The next time the smell of rain hangs heavy in the air, just wrinkle up your nose and tell your coworkers, "It's going to rain, I can smell the 1,10 dimethyl–9–decanol."

CLOUD PREDICTIONS

I bring fresh showers for the thirsting flowers,
From the seas and the streams;
I bear light shade for the leaves when laid,
In their noonday dreams.
From my wings are shaken the dews that waken,
The sweet buds every one,
When rocked to rest on their mother's breast,
As she dances about the sun.
I weld the flail of the lashing hail,
And whiten the green plains under,
And then again I dissolve it in rain,
And laugh as I pass in thunder."

—Percy Bysshe Shelley, "The Cloud"

 A cloud was a cloud until we developed a scientific way of thinking about it. The first person to try and distinguish clouds was the French scientist Jean Baptiste Lamarck, who is known for his theory of inherited acquired characteristics. In 1802 he listed 5 types of clouds which he expanded to twelve classifications in 1805. His listings are:

Hazy clouds: *en forme de voile*
Massed clouds: *attroupes*
Dappled clouds: *pommeles*
Broomlike clouds: *en balayeurs*
Grouped clouds: *groupes*

This classification was ignored by his countrymen, probably because it added little to the description and understanding of clouds. When Lamarck added weather predictions to his yearbook based on the moon, Napoleon was said to have taken him aside at an official reception and requested that he restrict his studies to "natural history."

The sailors classed the cloud sequence of a warm front as "cat tails" (cirrus) giving way to "flocks of sheep" (altocumulus) before developing into "cotton bales" (cumulus).

In 1802, a London pharmacist worked out a paper on "The Modifications of Clouds." Luke Howard used Latin names to describe the different types of clouds. He used the word for hair to describe cirrus, heap for cumulus, layer for stratus, and rain for nimbus to describe the clouds in the sky. Later he added alto and fracto, meaning middle and broken to describe the clouds.

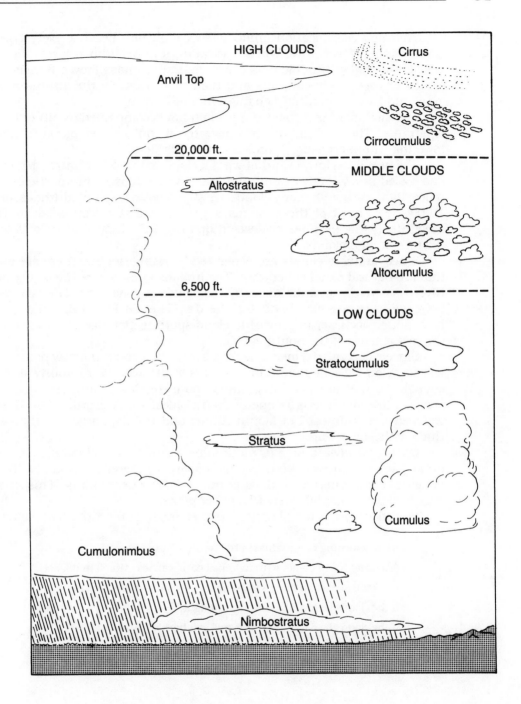

This new classification immediately caught on, because every type of cloud was self-descriptive. When a warm front moves into an area, the first sign is the high-altitude cirrus cloud. As the air mass moves further, the clouds change to altocumulus and then to stratus. As the stratus moves lower and lower it eventually changes to nimbostratus.

The movement of a cold front is often not accompanied by either cirrus or stratus. The cold air moves under the warm air and plows up giant towering "thunderheads," which are cumulus clouds.

The best descriptive word of what a cloud is inside, is "mist." Moisture has cooled below its dew point and has now condensed into particles that reflect light. Although many clouds are supercooled water at altitudes above 18,000 feet, most of these clouds are ice crystals and snowflakes. This accounts for the peculiar, iridescent quality of the light that reflects from the tops of high clouds.

Since cumulus clouds are generated by rising hot air, there are areas that are natural cloud generators. The Indians knew where the Mississippi River was even from a distance, by the clouds it generated. The islands of Polynesia generate clouds during the day. The old Polynesian navigators had enough accuracy to get within cloud-spotting distance as they sailed in the vast reaches of the South Pacific.

Since clouds are formed whenever air is lifted above the dew point, they will form on hills or mountains whenever the relative humidity is high enough, so that elevated air is lifted above the dew point. Seeing clouds form on low mountains is one of man's oldest signs of rain. The Greeks watched the "rain-cap" of Mount Althos and the Japanese did the same thing with Mount Fuji.

On the windward side of mountains a series of "roll clouds" is often formed at a distance of about five miles apart. Wherever the cloud is, the air is being lifted. Wherever it disappears, the air is descending. This knowledge is of fundamental interest to glider pilots.

Proverbial lore is full of sayings that express some truth about clouds:

> In the morning mountains, in the evening fountains.
>
> A round-topped cloud, with flattened base, carries rainfall in its face.
>
> Mackeral skies and mare's tails, make lofty ships carry low sails.
>
> If the clouds be bright, t'will clear tonight,
>
> If the clouds be dark, t'will rain, will you hark?

PROPHETIC SKIES

Red sky in the morning, sailors warning; red sky at night, sailors delight. (English)
Evening red and morning gray, you're sure to have a fishing day. (English)

Stay at home when the morning sky appears red, but look for a good day's travel when the evening clouds turn crimson. (Chinese)
When it is evening you say, "It will be fair weather, for the sky is red." And in the morning, "It will be stormy today for the sky is red and threatening." Matthew 16:2
Rouge soir et blanc matin, c'est le jour de pelerin. (French)
Temps rouge en abas, bone femme cherche du bois; temps rouge en amont, bonhomme couvre ta maison. (French)
Ponente rosso, levante adosso. (Italian)
Rossa la sera e bianco sul matteno, mettite allegro in viaggio, o pellegrine. (Italian)

ALL OF THESE PROVERBS SAY THE SAME THING.

Perhaps the oldest of all weather lore is the idea of the red sunset bringing a good following day. It is even more surprising that the same sign in the morning means bad weather. When you look at an evening sunset you are looking at up to four hundred miles of westward sky. Generally speaking, about seven out of ten red sunsets are predictive of good weather in the northern climates. The excessively red color of a sunrise that brings bad weather is due to thin cirrus clouds scattering the light at the edge of a warm front.

A fire-colored red sunset is occasionally seen up to twenty minutes after the sun has sunk below the horizon. This burning color is a good indication of a tropical storm. The low barometric pressures and strong winds carry clouds to extreme heights and they scatter light for four to six hundred miles.

On land a yellow sunset is often an indication of large amounts of dust in the sky. This is often seen in the desert areas when high winds are coming.

It is a popular belief that when the sun sets between cloud bands with a reddish yellow cast the storm will come from the southwest. If the sun rises with a reddish yellow cast the storm will come from the east. This may not be true in all areas, but there has been a study done of excessively red and yellow sunsets and sunrises. This study tabulated the percentage of time that rain followed within 24 hours of an unusually red or yellow sunset.

	Winter		Summer	
	Red	Yellow	Red	Yellow
Sunrise	61%	40%	70%	38%
Sunset	29%	58%	43%	51%

As your eyes travel downward to the setting sun, you can see the blue fading into faint green, and then into the yellow and red bands. The green coloration is a more accurate predictor of rain than the red or yellow, but it is subtle. I have often noticed a distinct green hue above a red sunset. The next day it would rain in spite of the red sunset. The sailors used to say, "A green sky for wind."

In winter time, a green sky is often an indication of snow. When the sky turns emerald green at sunrise, there is likely to be heavy snow that day. This weather sign is watched in the Alps.

The "green flash" is perhaps the most interesting of all sky phenomena. As the sun rises or sinks below the horizon, an island of yellow light seems to float above the sun. The yellow drains away and green rays flash across the horizon. This phenomenon has little to do with the weather, but high-altitude winds may increase its possibility.

A brown ring around the sun is associated with volcanic dust in the atmosphere. It is called the "Bishop's Ring" for it was first described by the Bishop of Honolulu. It is an indication of cooler temperatures and harder winters because more of the sun's rays are being scattered away from the earth. Generally, there is less rainfall when there is an excessive amount of volcanic dust in the atmosphere.

Along the coasts it is common to see a dark blue line on the horizon. Normally air at ground level is warmest, but in this case the cold layer of air is just above the ocean. This "blue horizon" can be a result of a cold front, but generally it is due to cold sea air.

Nearly all unusual weather imagery is due to thermal stratification of the air. A stick poking out of cold water with a layer of warm air above it can assume substantial magnification at a distance. We hope there will always be a monster hidden deep in the depths of Loch Ness where scientists can't reach. On a practical level, we recognize that most sightings are made in summer with warm air and cold waters producing mirage conditions.

Warm fronts can produce the thermal stratifications leading to mirages, but often they are a natural result of temperature combinations. Some of the unusual stories about mirages include the Navy pilot who found himself

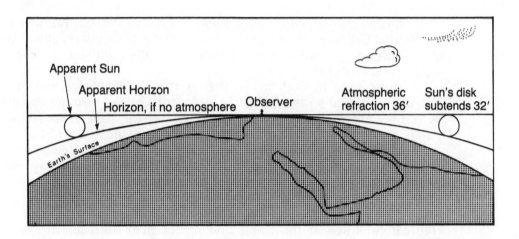

on a collision course with a ship 3,000 feet above the sea. Another story involves the Canadian fisherman who saw the city of Cleveland upside down while fishing eighty miles away. He could clearly distinguish the horses and buggies and streetcars.

A study done in Utah found that every type of mirage had its own pattern of thermal stratification. As we journey by car through the desert we can see the distant lake and the blurry waving palm trees far in the distance. These are formed with high temperatures and little wind. A mirror is formed from a ground temperature of about 125°F and a temperature of 95°F at five feet above ground level. This mirror reflects the blue sky. Hot air escaping from surface features provides the swaying trees.

A reversed form of this mirage forms over cold ground just before sunrise when the air is hot. The military police at the Dugway, Utah, base keep careful track of the traffic and are familiar with another strange mirage. It's always a surprise for them to count five pair of headlights on the distant road and then find that only one truck is pulling into the base.

The twinkling of stars is directly linked to increased relative humidity. Stars on the horizon up to about 40° of altitude will twinkle in normal weather, but the stars directly overhead don't twinkle. As the relative humidity increases, or high winds disturb the upper air, the altitude of star twinkling rises. In the tropics, people realize that high altitude twinkling stars are a sure sign that the rainy season is about to begin.

"When the stars begin to huddle, the earth turns into a puddle," is another statement of relative humidity. It means that more humidity blurrs out fainter stars. Blue light is a shorter wavelength, and travels farther through water than red light. When the stars twinkle green, this is a sign of good weather. But when they twinkle blue, it's a sign to get your raincoat out.

THE RISING OF THE MOUNTAINS

While the hush yet brooded, the messengers of the coming resurrection appeared in the East. A growing warmth suffused the horizon, and soon the sun emerged and looked out over the cloudwaste, flinging bars of ruddy light across it, staining its folds and billow-caps with blushes, purpling the shaded troughs between, and glorifying the massy vapor palaces and cathedrals with a wasteful splendor of all blendings and combinations of rich coloring. It was the sublimest spectacle I ever witnessed, and I think the memory of it will remain with me always.

—Mark Twain, *Roughing It*

If you live in a place with mountains, it is easy to see that the relative heights constantly change. On an early clear morning, the Olympic Mountains west of Seattle tower in the sky, but as the day progresses, they don't appear to be so high. In the late afternoon the mountains seem larger and they sink as the evening cools. These apparent changes in height are due to temperature stratification and changes in relative humidity.

A higher than normal appearance of distant hills and landmarks is an old weather sign, and many proverbs speak of the rise in height and increase in nearness before stormy weather arrives. "The further the sight, the nearer the rain," is a statement of practical observation. France is only a line on the horizon to an English observer standing on the white cliffs of Dover. On occasion, observers have been able to see the distant landmarks of France clearly, as a warm front raised the apparent height of the land.

The arctic mirage, which raises and magnifies distant objects, is known by Icelanders as the Hillingar Effect. In the northern climates, it is as common as the familiar lake mirage of the hot deserts, but it is due to the opposite form of temperature stratification.

During the cold clear nights, the ground radiates heat into the sky, and the overlaying air is warmer in relationship to the ground. The mountains appear higher and they gradually sink as the sun warms the overlying air. In the late afternoon, the same process causes the mountains to appear to rise and become more blurred in shape.

When a warm front moves into an area, the ground is cold in relationship to overlying air. The mountains rise in response to this temperature difference. If a cold front moves into an area, the mountains will sink until the temperature differences become normal.

Our view of the world is produced by a temperature drop of 1°F per 300 feet of elevation. If it is 80°F at ground level, then it will be 70°F at an

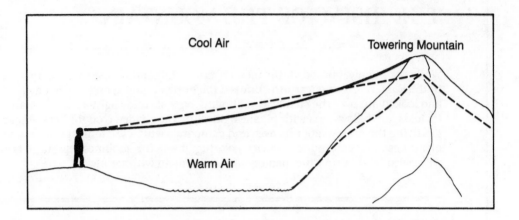

altitude of 3,000 feet above us. This temperature stratification produces a
gently curved view of the world and ships gradually sink below the horizon.
If instead of dropping, the temperature rises, then light bends differently
and we see a much different world. If the temperature becomes 18°F warmer
at a height of 300 feet, then the earth appears flat, and we can see at much
greater distances. If the temperature rises by 30°F in 300 feet, it bends
light, and we appear to be at the bottom of a giant saucer surrounded by
walls of the horizon in all directions. In the Arctic regions, the sea is very
cold from melting ice but the sun shines for twenty-four hours during the
summer. As a result, it creates a strong Hillinger Effect.

Summer in the Arctic becomes a magic fairyland. Distant objects are
raised and distorted by temperature differences. Admiral Perry believed that
he had discovered a vast new land northwest of Greenland. It was called
"Crocker land," but when the Eskimos were asked, they replied that it was
only "poojok," that is, "mist." It was simply ice flows raised and distorted.

The Asiatics had their mythological city of Shamballa in the glaciers of
the Himalayas. The Vikings had the land of Ultima Thule in the mysterious
Arctic. This was a land at the edge of the world where all waters flowed down
into the earth to replenish the springs and rivers. The world "Thule" is a
Celtic term meaning "to raise oneself." It appears to be an interesting trick
of nature.

The high temperature lapse rate that produces the saucer-shaped view
of the world has produced a strange body of "hollow earth" literature. It is
not surprising that the first writer on the hollow earth genre was a Norwegian fisherman. Hollow earth believers point to Admiral Byrd's expedition
where he visited a volcanic island west of Greenland where green vegetation
flourished amid the icebergs. This account, coupled with the Hillinger
Effect, produced the legends of the hollow earth.

"Ice blink" is another phenomenon in which the warm sky acts as a
mirror to reflect the imagery of the cold waters. Sailors in the northern

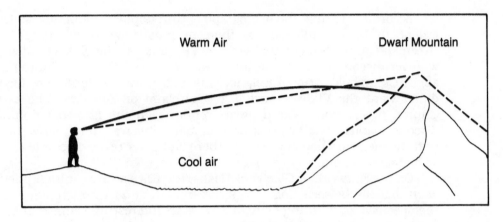

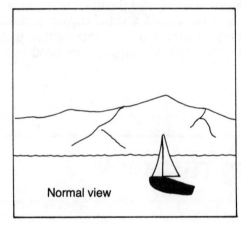

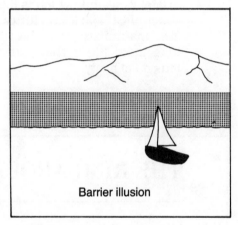

waters can see icebergs and ice leads by looking at the sky. In these dangerous waters, this is a good way of finding passage through the ice pack.

Students of history have wondered what motivated the early Vikings to sail hundreds of miles through rough waters with crude navigational equipment. The early sailors had to wrestle with accounts of whirlpools, sea monsters, sea fences, and possibly the edge of the earth. It is quite possible that the arctic mirage enabled them to see the distant lands.

In 1939, Captain Robert Bartles was able to see the Snaefell Jokull volcano on the west coast of Iceland from a distance of three hundred miles. Although Iceland would be about forty miles below the line of sight from the Faeroe Islands, perhaps it was occasionally seen. Greenland, too, could have been seen from Iceland, given the temperature stratification so the brave Vikings may have had reasons for their voyages.

According to his son's biography, Christopher Columbus sailed to Iceland in 1477. While he was there he must have heard stories about Greenland and possibly the saga of Vineland. Medieval law states that whoever discovered land first laid claim to it. The Catholic Church knew about Columbus's trip to Iceland, which they wanted forgotten, for they didn't want the Vikings having a legal claim on America. The earliest Church descriptions placed great emphasis on the fact that Columbus discoverd lands never before known to man. This was to reinforce the idea that the men who had always lived there had never really discovered it, and so had no real claim.

On the Fairweather Glacier of Alaska, there is a strange mirage phenomenon that can be seen from 7 to 9 P.M. from June 21 through July 10. It is the spectacle of a strange city. There is an interesting claim that this is Bristol, England, which is 6,000 miles away, and the mast of a ship and the Tower of St. Mary Redcliff can be seen in this distant corner of the earth. It seems unlikely that the Hillengar Effect could operate at this distance. It is more likely that natural geographic features are producing the illusion of a city.

THE RING AROUND THE MOON

Then up and spake an old sailor, had sailed to the Spanish Main, "I pray thee, put into yonder port, for I fear a hurricane. Last night the moon has a golden ring, and tonight no moon we see!" The skipper, he blew a whiff from his pipe, and a scornful laugh laughed he. Colder and louder blew the wind, a gale from the Northeast. The snow fell hissing in the brine, and the billows frothed like yeast. Down came the storm, and smote amain, the vessel in its strength.
    —HENRY WADSWORTH LONGFELLOW, "THE WRECK OF THE HESPERUS"

The oldest known sailor's song is the "Ballad of Sir Patrick Spence," which goes back about eight hundred years to a time when England had no navy. The king wanted to do a little privateering to earn money for the royal treasury. He orders the best English sailor, who was Patrick Spence, to serve as his captain. The crew sailed at a time when the "new moon was holding the old moon in its arms." A storm arose, the crew lost their lives, and the belief that an odd lunar ring was a storm sign

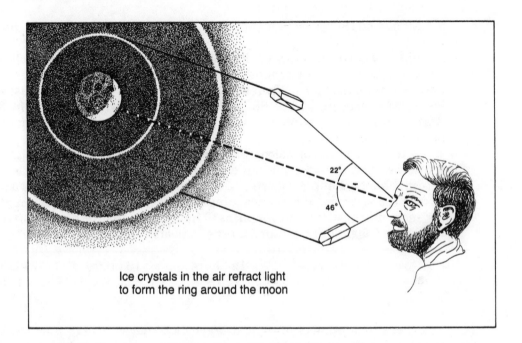

Ice crystals in the air refract light
to form the ring around the moon

was strengthened. This is not a true lunar ring, but a lunar crescent around the old moon. It is usually seen in the spring or late fall just after sunset, or before sunrise.

There are several different-sized rings that can form around either the sun or the moon. Rings are much more common around the sun, but we notice the ring around the moon more often. The rings are formed by the scattering of light by ice crystals in the atmosphere. They usually form at the edges of storm centers, so they do have predictive value.

The largest common ring forms at a 46° radius to the moon. It is at the distant edge of storm centers and means that the storm is twenty-four to thirty-six hours away. The old proverb refers to the 22° and 46° haloes this way: "Circle near, water far; circle far, water near." The proverb should correctly read: "Circle far, water far; circle near, water near."

The most common ring forms at a 22° radius to the moon. It generally means that a storm is twelve to eighteen hours away. But it could be an hour away and it could miss the area completely. Both rings are formed by the same crystals, but in different orientations.

Your chances of accurately predicting rain depend upon your location. A halo in Fort Worth, Texas means that there is a 36 percent chance of rain in the next day. A halo in Columbus, Ohio indicates a 65 percent chance of rain in 24 hours. The closer you are to normal storm tracks, the more likely the halo predicts rain. During the summer a halo means that the storm is an average of 18 hours away, but during the winter it is about 11 hours away.

When the barometer was falling 83 percent of Ohio haloes were followed by rain in 24 hours. On a rising or low barometer only 53 percent of haloes were forerunners of rain. When winds were blowing from the southeast and a ring was seen, rain came 88 percent of the time within 2 days. In Fort Worth, Texas an east wind with a halo produced rain or snow in 2 days 95 percent of the time.

A ten-year study of haloes was done in Pennsylvania, with respect to future weather. On a day without either a lunar or solar halo there was a 50 percent chance that the following 2 days would have rain. If there was a halo during the summer, the chance jumped to 77 percent, and to 87 percent for snow during the winter months. The average chance of precipitation was plotted against the seasons for the time after a halo.

	6 hours	12 hours	18 hours	24 hours
Summer	18%	25%	31%	42%
Winter	27%	45%	66%	78%

There are a number of superstitions about haloes. There is no truth to the idea that the number of days before a storm is indicated by the number of stars trapped in the lunar halo. An open side to the lunar halo is said to indicate the direction of the storm, but this isn't necessarily true.

The same ice crystals that produce rings, can produce unusual effects at low sun angles. Arcs, mock suns and sun pillars are variations of the same ice crystals. Jacob's ladder and Ezekiel's wheel's would be of interest to the student of haloes. The signs in the sky that once announced the birth of kings and saviors in ancient times seem to have been haloes.

Other special signs from God were meteorites and northern lights. When Hannibal brought his elephants to Italy, there was a shower of hot stones with thunder and lightning. A scroll appeared with the words, "Mars himself stirs his arms." It was a good moral boost for an army facing a tough war. Josephus wrote that prior to the Jewish revolt of A.D. 66, chariots and armed battalions were seen in the sky. These were probably the northern lights which are uncommon in southern Europe.

THE CROSS IN THE SKY

On the twenty-first of January, 1671, one or two hours before sunset, the first Parhelion was seen at the Bay des Puans. High in the air was seen a great crescent, its horns pointing heavenward; while on the two sides of the sun were two other suns, at equal distances from the real one, which occupied the middle. It is true, they were not entirely revealed, as they were covered in part by a rainbow-hued cloud, and in part by an intense white radiance, which prevented the eye from clearly distinguishing them. When the savages saw this, they said that it was the sign of a severe cold spell; and indeed the succeeding days were extremely cold.

—"THE JESUIT RELATIONS"

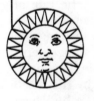

On October 28, A.D. 312, the Battle of Milvian Bridge changed the destiny of the Roman Empire. Constantine defeated his rival Maxentius, and Christianity became the religion of the Empire. The story of the miraculous cross in the sky paved the way for the change.

Christianity underwent a series of persecutions, and only 7 years earlier Diocletian tried to wipe out all Christians. Historians believe that only about 5 percent of the Roman people considered themselves Christians at this time. The father of Constantine took the title of "Sol Invictus" meaning "The Unconquered Sun," because sun worship was common. Constantine had been a follower of the Greek god Apollo before the battle.

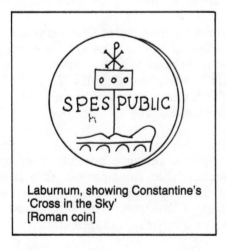

SPES PUBLIC

Laburnum, showing Constantine's
'Cross in the Sky'
[Roman coin]

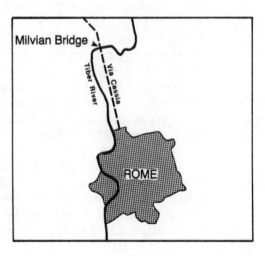

Milvian Bridge

Tiber River

Via Cassia

ROME

Battle in ancient times was more traumatic than today for there was no honorable surrender. The defeated army might be killed as rebels or paraded in Rome and sold into slavery. Before battle it was considered important to look at omens as signs from the gods. Notes were taken of animal livers, the flight of birds, and lightning as an indication of success.

The evening before the Battle of Milvian Bridge, a sign appeared in the sky which aroused the attention of the entire army. Eusebius, an influential Christian in Constantine's court, gave the first account in "Ecclesiastical History." He says nothing about a vision, but that Christ gave Constantine help before the battle.

An unknown writer put out a booklet entitled "On the Deaths of the Persecuters." It says that Constantine had a dream before the decisive battle, and as a result he inscribed the heavenly sign of the Christian God on his battle shields.

Another writer described both a dream and a vision in "The Life of Constantine." "When the day was beginning to decline, he saw with his own eyes the trophy of a cross, light in the heavens above the sun bearing the inscription "In this conquer." At this sight he was struck with amazement, as was his whole army, which witnessed the miracle. And while he continued to ponder its meaning night suddenly came on; then in his sleep the Christ of God appeared to him with the same sign which he had seen in the heavens."

The early Christians used the sign of the cross and the fish to represent their religion. There appears to be little doubt that the cross of Christ was T shaped. The earliest Christian cross was found above a little stone altar in the buried city of Pompeio less than 10 years after St. Paul had visited it.

The writer of "The Life of Constantine" tells us, "Constantine was admonished in a dream to inscribe on the shields the heavenly sign of God and thus to commit himself to battle. He obeyed and inscribed the sign of Christ on the shields; the Greek letter "X" intersected with the letter "I" bent at the top . . ."

By the year A.D. 317 coins bearing the shield of Constantine's army began to appear. It is likely that some of the designs on the coins were inscribed by army veterans who had been present at the battle.

According to historians, there is little evidence that the "X" and "I" had been used as a Christian symbol. The X was a Celtic symbol of good fortune and became known as "St. Andrew's Cross." The "X" was the Greek letter for Ch, the first letter of Christ.

Constantine's vision has been considered to be the result of a dream, but the fact is his army saw it. In spite of previous hostility, they readily became Christians, so it had to be more than imagination. Fritz Heiland, the German astronomer, believed that the cross was a conjunction of Saturn, Mars, and Jupiter in the sign of Capricorn. This might have been of interest, but it would hardly produce a sense of "amazement."

A study of the army shields on Roman coins shows us what probably occurred. The vertical "I" through the sun is a phenomenon generally known as a "sun pillar." It is produced by needles of ice crystals falling through still air. The two dots on either side of the "X" are mock suns, although many coins have 3 circles representing the 3 suns below the Christogram. The "X" in the sky is the result of an encircled sun with an arc above it. This is uncommon, but it happens as a result of a cold front with ice crystals falling from cirrus clouds at a low sun angle. The interpretation of the vision, "In this Sign Conquer," probably was the result of a dream.

Constantine was a genius as a political leader. He was honest and fair, and the changes he made endeared him to the Roman people. He did not act hastily, but by a series of laws gradually turned the Roman Empire into a Christian empire. His most famous edict read, "All the judges, the city folk and the craftsmen shall cease work on the day of the sun. However the farmers may continue their labors in the fields, because often no day is more suitable for sowing and planting . . . This was the political origin of Sunday.

To an army looking for a sign of divine will, it was a miracle; to a meteorologist, it is an uncommon case of the refraction of light through plate and needle crystals of ice. Taken as human or divine, a cold front produced the sign that changed history.

THE TURN OF THE TIDE

When first the moon appears, if then she shrouds,
Her silver crescent, tipped with sable clouds,
Conclude she bodes a tempest on the main,
And brews for fields impetuous floods of rain.
Or if her face with fiery flushings glow,
Expect the rattling wind aloft to blow;
But four nights old (for what is the best sign),
With sharpened horns, if glorious then she shine,
Next day not only that, but all the moon,
Till her revolving race be wholly run,
Are void of tempests both by land and sea.

—AUTHOR UNKNOWN

It was not obvious to early man that the moon was the cause of the tides. Even Galileo was angry with Kepler for "Having given his ear and consent to the moon's predominancy over the water and to occult properties and suchlike trifles." Isaac Newton established the laws of gravitational attraction in the "Principia". He found that the height of the tides were due to the distance of the earth to the moon and the alignment of the sun and the moon.

The ocean tides rise up to 50 feet from the pull of the moon. Even the solid earth rises by 4 to 40 inches as the moon passes overhead. Many springs and wells increase their level of water, but occasionally they fall when the moon is overhead. The moon is lifting the earth and thus lowering the level of water.

Robert Boyle was the first to use a barometer to try and measure lunar tides. He wasn't able to find any air tides as a result of the moon but he did establish the relationship between storms and barometer readings. Since there are so many irregularities it wasn't until 1842 that lunar atmospheric

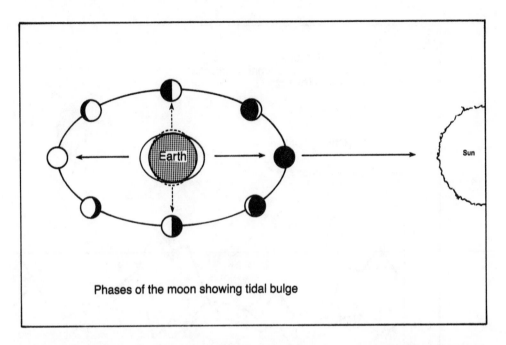

Phases of the moon showing tidal bulge

tides were able to be recorded. While there are no true atmospheric tides, there is an ionospheric lunar tide. When this lunar tide occurs at the same time as the sunset, it becomes more brilliant.

It has been more than a hundred years ago since German meteorologist noted that thunderstorms usually come with the hour of the high tide in the North Sea, the Bay of Bengal, and along the American coast. The relationship between the tides and rain, or changes in wind, has been commented on by sailors and fishermen along the English and the New England coast.

The strange behavior of the weather at the turn of the tide resulted in these old New England proverbs:

> If it raineth at tide's flow,
> You may go and mow:
> But if it raineth at the ebb,
> Then if you like, go off to bed.

> Rain on the flood, only a scud,
> Rain on the ebb, sailors to bed.

Odd tidal phenomena has been the subject of a number of observations by New England sailors. In Delaware Bay, the incoming tide appears to carry storms to the north, and the outgoing tides seem to direct them to the south. When the tides are slack, the storms move straight eastward.

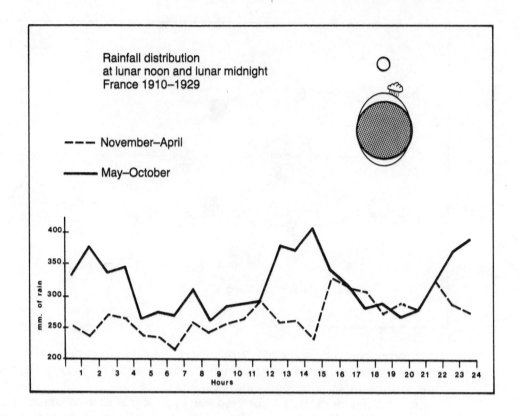

Rainfall distribution
at lunar noon and lunar midnight
France 1910–1929

– – – November–April

——— May–October

As the tide rises toward its peak, the air may be calm, but at the point of high tide a strong southwest wind picks up. This wind generally comes on calm days with little wind between the hours of 8:00 A.M. and 3:00 P.M. The strongest tidal winds are in the spring, and they come 50 minutes later each day.

If the winds are already in the southwest, and the sky is cloudy, the odds are very good that the rain will come with high tide. If the time passes without rain, the rain is likely to begin at the next high tide 12½ hours later. This may depend on the time of the year, but it has been confirmed by many observations.

When French meteorologists plotted high tide as "lunar noon" and graphed the rainfall distribution against this, rain usually occurred at lunar noon or lunar 1:00 P.M. During the winter months it is less likely to rain or snow during these tides. The U.S. Weather Bureau has also studied this, and they found a complex interaction between the most likely times for rain during the solar day and lunar tides.

Twice each month, the moon lines up with the sun, and produces the highest tides. The moon travels in an elliptical orbit, so its low point does not normally line up with the sun. When the perigee (low orbital point) lines up with the sun and the earth, very high tides are to be expected.

Fergus Wood used a computer at the U.S. Naval Observatory to determine all of the highest tides in historical times. Then he went back through old newspapers and weather records to see if anything unusual had happened. His earliest example was August 4, 1635, when Governor William Bradford of Plymouth, Massachusetts, wrote "Such a mighty storm of wind and rain as none living in these parts either English or Indians ever saw. . ." Bradford mentioned that an eclipse had occurred two days earlier, meaning that a very close lineup of the moon with the sun had taken place.

Wood dug up many cases in which lunar tides brought high waves and winds. During winter, very high tides tend to produce blizzards and misery. But often there was a high pressure area over the coast and nothing unusual happened.

His computer printout extends far into the future, and there are a number of dates in the 1990s that would make a wise ocean dweller head for the real-estate office. Although it cannot be said that there will be severe storms on those dates, if they do occur, there will be major damage.

We do not feel the tides, but evidence suggests that much of nature does. A century ago, William Ross kept a salt water aquarium in his home in Devon, England. During the high tide a shanny (Blennis pholis) would swim in the shallow water above a big rock. As low tide came it would shift to the deeper water in the aquarium. There was no movement of water and nothing to indicate what was happening outside, but the fish knew the rise and fall of the tides by instinct.

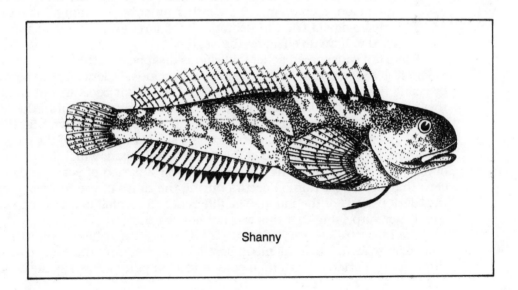

Shanny

Some evidence suggests that our subtle functions feel the tides. The tide provides a weak biorhythm that may influence our behavior. Like the weather it is subtle and changeable.

THE MOON AND THE WEATHER

Now the weather depends on the moon as a rule,
And I've found that the saying is true.
In Glen Nevis, it rains when the moon's at the "full."
And it rains when the moon's at the "new."
When the moon's at the "quarter," then down comes the rain;
At the "half's it's no better, I ween;
When the moon's at "three-quarter's" it's at it again,
And it rains besides mostly between.
                              —Old English poem on the relationship of the moon
                                           to the weather, author unknown.

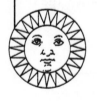

In early times man must have looked out of his cave and thought "the moon is changing, the weather must be changing." The early meteorologists tried to debunk all folklore, and the moon was believed to have no effect on the weather.

There is a widespread belief that clouds disappear at the time of the full moon. It is believed that "the full moon eats clouds." Clouds are generated by rising hot air in the late afternoon, and as the air cools in the evening these hot air clouds disappear. We don't normally notice this until the time of the full moon, when we can see it. Several studies have shown that there no real difference in night cloudiness of the sky at any phase of the moon.

We notice the lunar period of full moon to full moon, which takes 29.53 days. When rainfall patterns are averaged over a period of years, it can be seen that the lowest rainfall occurs during the times of the full moon and the "dark period" of the moon. The difference in rainfall is only 5 percent, so it is not surprising that this was not noticed earlier.

The moon travels a full circle in 27.32 days, but it has to rotate for 2 more days to return to a full moon position, as seen from the earth. During this rotation period it has a high point and a low point known as the apogee

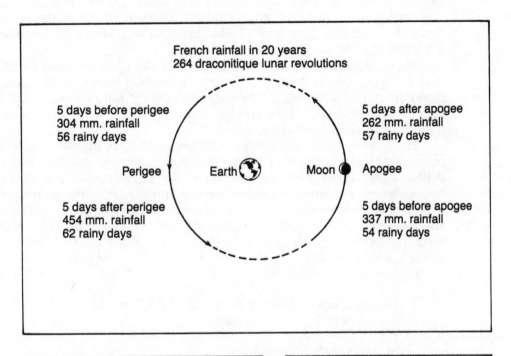

French rainfall in 20 years
264 draconitique lunar revolutions

5 days before perigee
304 mm. rainfall
56 rainy days

5 days after apogee
262 mm. rainfall
57 rainy days

Perigee   Earth   Moon   Apogee

5 days after perigee
454 mm. rainfall
62 rainy days

5 days before apogee
337 mm. rainfall
54 rainy days

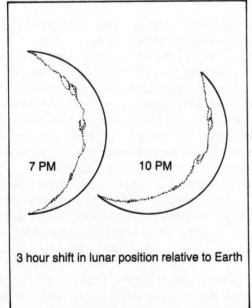

7 PM   10 PM

3 hour shift in lunar position relative to Earth

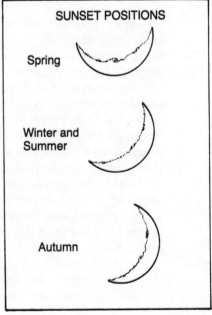

SUNSET POSITIONS

Spring

Winter and
Summer

Autumn

and the perigee. The perigee generates the highest tides, and theoretically should influence the weather more. The perigee does not influence the number of storms appreciably, but it does influence the amount of rain that falls at that time. A French study of the tidal effects of the moon is illustrated. Rainfall data from Alaska has also been shown to increase at the times of perigee.

Much folklore deals with the position of the quarter moon. A moon that is "lying on its back," obviously is holding water. If the quarter moon is standing upright, the water is spilling out. Occasionally, the beliefs are reversed, a standing moon is a dry moon, because it has already spilled its water.

The quarter moon is rotating daily about 360° in the sky. Since it is rotating 15° per hour, in 4 hours it will change from holding water to spilling water. We usually view the quarter moon after sunset, but if we look in the eastern sky in the morning, we will find that it is like an umbrella. Viewed just after sunset, the quarter moon holds water in spring and summer, and stands upright during the fall months. The belief that a standing moon is "spilling water" must have started in the Middle East where the fall rains are seasonal.

A Greek poem reads:

> When the moon lies on her back [spring],
> Then the southwest wind will crack.
> When she rises up and nods [summer]
> Then northeasters dry up the sods.

Luke Howard, the father of cloud names, was the first person to investigate the influence of the moon's declination on weather cycles. He published several papers around 1840 about lunar influences on long-term weather patterns. As we view it the moon's orbit is 5° higher in the sky or 5° lower compared to the earth's orbit over a 18.6 year pattern. If we look at the height of the mid-sky moon and then compare its position 9 years later, there will be as much as 10° difference. In 1901, a paper was published in Australia blaming the long droughts on the position of the moon. As the moon went southward, there were six years of moderate rains, but when the moon turned northward, there was a 12-year dry period.

French meteorologists have verified a similar effect in Europe. When the moon was at its northern declination, there were 840 days of rain compared to 775 days of rain in the cycle of southern declination. When the moon crosses the plane of the ecliptic from north to south, more rainfall occurs at the first quarter. If it crosses in the other direction, rainfall is enhanced at the third quarter.

Researchers have studied the 22-year sunspot cycle in relationship to rainfall, but the 18.6-year lunar declination cycle might play an equally

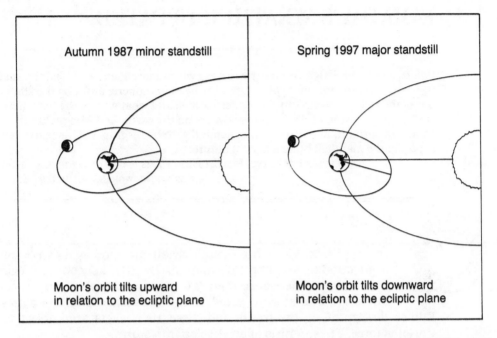

Autumn 1987 minor standstill

Moon's orbit tilts upward
in relation to the ecliptic plane

Spring 1997 major standstill

Moon's orbit tilts downward
in relation to the ecliptic plane

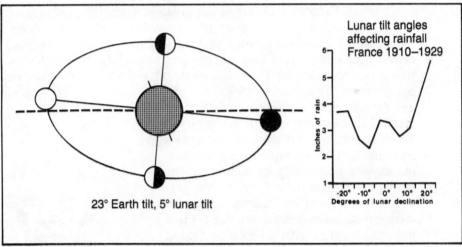

23° Earth tilt, 5° lunar tilt

Lunar tilt angles
affecting rainfall
France 1910–1929

Inches of rain

Degrees of lunar declination

important factor. There is a drought cycle of about 20 years on the Great Plains of the United States. This cycle appears to be a combination of the dry phases of the sunspot and lunar declination cycles.

Researchers have ignored the moon, because of the complexity of the many factors involved. Poets have always known that the moon influences love, and now meteorologists are certain that the moon plays an important part in the weather.

FARADAY'S WEATHER UPDATED

Before this he investigated how the roaring winds moved, and troubled the smooth waters of the sea and what spirit turns the heavenly sphere; and why the stars rise out of the red east to sink in the western waves; and what warms the lusty hours of early summer, which calls forth rosy flowers on the earth; and who makes the generous autumn, in a fruitful year, abound with heavy grapes. This man used to expound the manifold hidden laws of nature.

—("DE CONSOLATIONE PHILOSOPHIE.") THIS IS ONE OF GEOFFREY CHAUCER'S LESSER-KNOWN WORKS, INSPIRED BY BOETHIUS.

During World War I the Scandinavian meteorologists were cut off from outside weather information. By 1917, Jacob and Wilhelm Bjerkness had developed their "air mass" and "polar front" theories. The origins of storms were "fronts" between colliding air masses. Eddies develop along the fronts and these air pockets are subject to the Coriolus force. These whirls of air develop into storms.

Michael Faraday was one of the greatest scientists, although he is not associated with any one important theory. His work in chemistry laid the foundation for analytical work and the understanding of compounds. His observations on current flow laid the groundwork for the electric motor and the generator. He did all of his work on electricity 50 years before the electron was discovered in 1898.

Faraday wrote a book called, *Experimental Researches in Electricity* in 1850. In volume 3 he said, "It becomes a fair question of principle to inquire how far masses of air may be moved by the power of the magnetic force which pervades them. When two bulbs of oxygen in different states of density are subject to a powerful magnet with an intense field of force, the mechanical displacement of one by the other is more striking. We may well conceive that the enormous sum of oxygen present in only a few miles of heated or cooled atmosphere can compensate for the great difference of magnetic force, and so any change of place can cause currents of winds having their origin in magnetic power."

Michael Faraday

Each atom contains unpaired electrons which are normally, randomly oriented. When they line up, they form "domains," and the spinning electrons are the cause of magnetism. In 1821, Faraday discovered that the current in a magnetic field rotates, and this fact is the basis of the generator and the electric motor.

If an iron filing is placed 6 inches from a magnet nothing happens. But if a string of iron filings lead from the magnet, the force is conducted at a much greater distance. Each square yard of the earth has 19,000 pounds of air above it. The magnetic pull on oxygen may be small, but millions of tons of oxygen are involved.

Faraday's theory explains several puzzling facts. Storm centers follow long parabolic arcs with the center at the north magnetic pole, which is hundreds of miles from the North Pole. If the Coriolus force were the only rotary mechanism, then the center of the arc would be the North Pole. When there is increased sunspot activity, storm tracks across the United States shift northward. This may be explained by the magnetic oxygen attraction.

Climatologists have long puzzled over the fact that climates shift from wet forests to dry deserts. Over a period of hundreds of millions of years the continents have drifted, so this explains some climactic changes. The north magnetic pole seems to be drifting several miles per year. If Faraday's postulate is correct this shifting of storm tracks, could turn grasslands into deserts. This has happened within recent geological times in both North Africa and the western United States.

Faraday didn't know about ions, for they weren't discovered until the early part of the twentieth century. The lifting of air in a storm center creates a giant "pancake" of negatively charged air. Above this layer positive ions dominate all the way up to the jet stream at about 40,000 feet. The prevailing winds of our northern latitudes will carry these ions through the earth's magnetic field which is approximately ½ gauss.

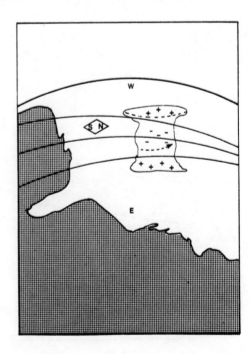

The direction and turning radius of electrons in the earth's magnetic field may be calculated by the formula used for cyclotrons. The stronger the magnetic field, the sharper the turning rate of the electrons. When this formula is used with the earth's magnetic field, it roughly describes the radius of a thunderstorm. The Coriolus force does provide an explanation for the swirling motion of air masses, but it does not explain why the winds in the southern quadrant should be so high. It is in this quadrant that the separation of ions takes place, and wind velocities are considerably weaker in the northern quadrant where few ions are generated.

Faraday also invented a remarkable motor which only recently has been studied and produced. Since it is not applicable to alternating current, it can only be used with direct current. If we visualize a cumulus cloud as a giant Faraday motor, with a current flow through the earth's magnetic field, then the powerful rotation effects in the southern quadrant of storms become obvious. Faraday did not apply his motor experiments to storms.

Why should earthquakes generate high winds and storms? The shifting of the earth does not shift air masses, but the squeezing of rocks between the earth's plates does release ions. These ions would develop a rotatory motion in the earth's magnetic field.

Why do displays of the northern lights tend to generate southerly winds and storms in certain locations? The answer may be in the rotatory action of ions which the sun sends into the earth's field. Astrophysicists speak of the twists in the northern lights as "Faraday rotation."

The tornado has defied the best attempts of scientists to understand it. One person who survived after getting a direct look into a tornado funnel, reported that the inside was like a cone of flame. This intense electrical

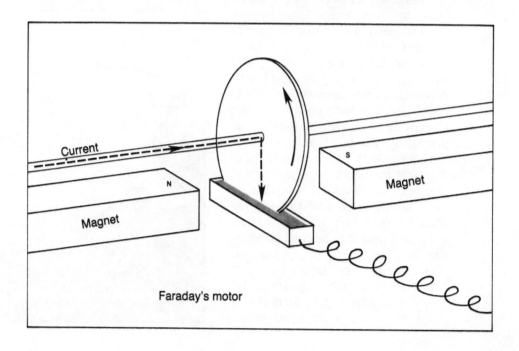

Faraday's motor

activity may be a Faraday motor. Since the flow of electrons at the leading edge of storms is usually towards the earth, this accounts for the fact that 99 percent of tornadoes rotate counterclockwise.

If a tornado is a Faraday motor operating by the earth's magnetic field and a current provided by storm clouds, then we have the possibility of controlling it. Scientists regularly send small rockets into clouds with wires attached to trigger artificial lightning. If we could do the same thing with tornadoes, then we could short circuit them and the driving energy would be gone.

Michael Faraday remains forgotten as the nineteenth century's last great genius of science without a formal education. The foundations of the electrical and chemical industries came about through his experiments. If you ask the name of the one scientist whose picture hung over Albert Einstein's desk, you will receive a surprising answer. It was Michael Faraday.

SOLAR WEATHER

Above the rest, the sun who never lies,
Foretells the change of weather in the skies;
For if he rise unwilling to his race,
Clouds on his brow and spots upon his face.
Or if through mists he shoots his sullen beams,
Frugal of light in loose and straggling streams,
Suspect a drizzling day and southern rain,
Fatal to fruits, and flocks and promised grain.

—AUTHOR UNKNOWN

It is believed that there were no planets around our sun 5 billion years ago. Scientists theorize that there was a near collision of our sun with another star 4.6 billion years ago. Streams of matter passed between the two stars as they sped apart. The forces of gravity began to coalese the heavier matter into planets.

If there were no sun, there would be no weather. Our planet would only be a frozen ball of ice without the changes in climate that has produced such diversity through the evolution of the DNA molecules.

The old Chinese were the first to notice sun spots. Dust storms left the sky so dark, that they were able to look directly at the sun during the day and the larger spots could be seen. When there were "imperfections" in the sun, these people believed they were due to the conduct of the Emperor and his court.

The eleven-year sunspot cycle has been intensively studied for more than a century. This cycle lasts from 7 to 17 years and averages 11.1 years. Since the magnetic polarity of the spots is reversed during the eleven years, it is actually a 22.2 year sunspot cycle.

Less carbon 14 is formed during the times of low sunspot activity. During the times of high solar activity more nitrates are formed in the earth's atmosphere, and their concentration has been studied for thousands of years from the Greenland ice cores. Scientists have also plotted the sunspot activity for the last 7,000 years from studying tree rings.

There were no sun spots during the years 1645 to 1715. During eclipses the sun's corona was dull and reddish. This 70-year period is known as the "Little Ice Age." Glaciers grew in size, farmers had a difficult time in northern latitudes, and winters were very severe. Were the earth's Ice Ages due to a lack of sun spots?

The weather varies with solar activity. When there are more sun spots, there are higher temperatures, longer growing seasons and more displays of the northern lights. High sunspot activity is also associated with suicides and war. Solar activity seems to influence the restlessness of man.

As we see it, the sun is rotating on its axis every 27 days. On the corona of the sun, giant flares spray ions into the surrounding space. The sun is divided into 4 quadrants of opposite magnetic signs. Every 7 days the magnetic sign changes. On the day the charged particles reaching earth change their sign, storm systems move further south and carry more moisture. The size of the low-pressure areas in the Gulf of Alaska, which often makes stormy weather across the United States, is directly related to the solar quadrant facing the earth.

The winter rains of California follow a regular 27-day cycle. The first autumn rain with (a precipitation level of) more than a quarter of an inch, begins the cycle. The rain cycle is divided into a 13-day wet half followed by a dry half. During the winter of 1972–73 there were seven 27-day cycles. The dry half of the cycles had 7 rainy days, and the wet half had 28 wet days.

Ronald Rosenberg and Paul Coleman discovered this cycle when they studied computerized rainfall records for Los Angeles during the last 50 years. They believe that during the winter months the polar air extends further south and affects California, but it has no effect during the summer. The sun drives storm systems across California like a gear with 4 cogs.

The sun is surrounded by a corona of ionized gasses. This is observed by an instrument that blocks out the disk of the sun, and allows scientists to observe the edges of the sun. This light can be broken down into a spectrum by a prism. The most interesting lines in this spectrum, for

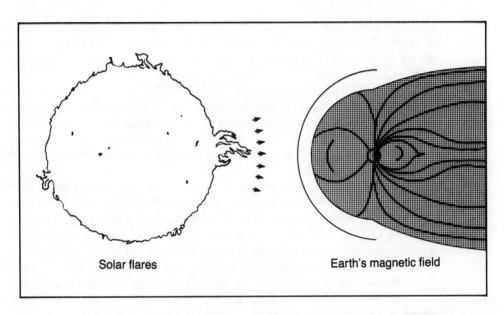

Solar flares   Earth's magnetic field

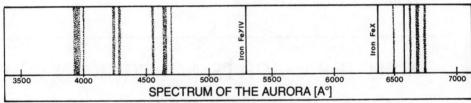

Iron FeXIV   Iron FeX

3500   4000   4500   5000   5500   6000   6500   7000

SPECTRUM OF THE AURORA [A°]

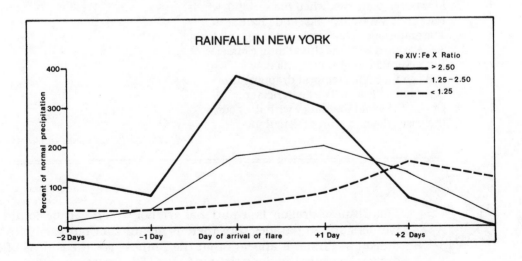

RAINFALL IN NEW YORK

observing the high energy activity of the corona, are the red and green lines of iron. When iron loses 10 electrons, it emits red light with a length of 6,174 angstroms. If it loses 14 electrons, it emits higher energy green light at 5,303 angstroms.

When there are flares and greater solar activity, the green line increases. The ratio of the red to the green iron line is an excellent index to future earth weather. Where the increased green iron light faces the earth, temperatures increase. The warmest day normally follows 2 days after the maximum of green iron light, and this is generally followed by colder temperatures for the next 5 days.

Unfortunately, these means of using the sun to predict the weather are far from the hands of amateurs, and even specialists have problems with this method, interesting as it is. It is known that the sun emits a great deal of noise at 167, 200, and 460 megahertz. Could the ratios of these wavelengths provide us with a look at forthcoming weather?

There is an old belief that configurations of the planets in space raise tides on the sun and thus create flares and sunspots. Should it prove to be true and better understood, we might be able to predict the direction of the weather for hundreds of years.

NORTHERN LIGHTS, SOUTHERN WINDS

I have seen tempests, when the scolding winds,
Have riv'd the knotty oaks; and I have seen,
The ambitious ocean swell and rage, and foam,
To be exalted with the threat'ning clouds:
But never till tonight, never till now,
Did I go through a tempest dropping fire.
Either there is a civil strife in heaven.
Or else the world, too saucy with the gods,
Increases them to send destruction.

—WILLIAM SHAKESPEARE, *JULIUS CAESAR*

The Chinese dragon is a national symbol which can be found on banners and in parades. But there is no animal that resembles the dragon of China. It appears that the Chinese got their ideas of the dragon by watching the northern lights. The long, snake like writhing lights inspired their beliefs.

There appears to be a belief even in ancient times that an aurora in the sky was a sign of colder weather, as this poem by the Roman poet Virgil shows:

> If Aurora with half open eyes,
> And a pale sickly cheek salutes the skies,
> How shall the vines with tender leaves defend,
> Her teaming clusters when the storms descend.

The first person to study the relationship of the aurora borealis and the weather was the pioneer English chemist John Dalton. He kept track of all the days that the aurora was visible from his home in England, and what the weather was like the following day. He found that there was no real difference in frequency between normal English rainy weather, and the weather the day after the aurora was visible.

Most folklore traditions believe that bad weather follows a bright display of the northern lights. The inhabitants of the Faroe Islands, which are north of England, believe that a low-level display means good weather, but a very active display means bad weather. In both Norway and the Faroes an "uneasy" display of the lights means windy weather. In Sweden it is believed that an early display of the northern lights means a severe winter. The Creek and the Cheyenne Indians of the United States looked for bad weather after an aural display.

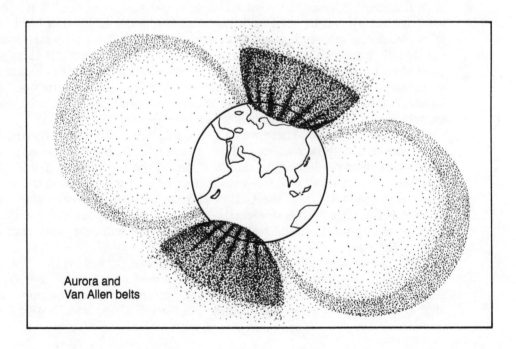

Aurora and
Van Allen belts

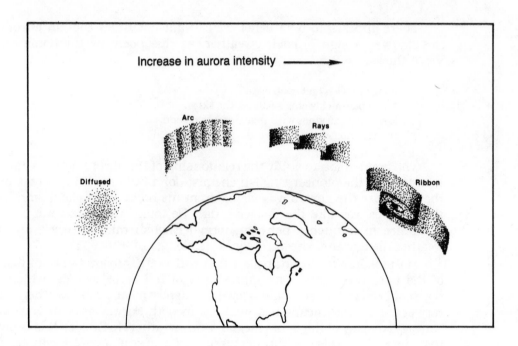

Increase in aurora intensity ———▶

Arc

Rays

Diffused

Ribbon

Most meterologists believe that the northern lights have nothing to do with the weather because they can be seen every night in the far north. A low-level aurora can be seen each night with blueish and yellow light. But what about real aurora displays, with deep red and brilliant pink streamers?

The first modern hint that scientists should watch the night skies was published in 1957. An airline weather forecaster noted that his computer predictions were wrong 10 to 14 days after a vigorous show of the northern lights. A study of Colorado weather charts revealed that there was major switch to rain and snow 13 days after an intense display.

On April 27, 1955, there was a great magnetic storm that turned the night sky into a show rivaling July 4. There had been a real drought and forecasters were predicting no rain in sight on the morning of May 9. That evening, the prevailing westerly winds stopped and cold air from the north met warm moist air from the Gulf of Mexico. During the day of May 10, a good rainfall soaked the dry earth. John Dalton was wrong: he didn't look far enough ahead, and he didn't separate the weak displays from the strong displays.

The oldest scientific report on the aurora was published in 1774 and addressed to Benjamin Franklin. Franklin followed with a short note saying that these aurora observations were of real interest. The observations were that severe auroras generally produce southwest winds within 48 hours.

English sea captains recognized this, and if there were other signs of bad weather they avoided sailing on the English Channel at that time.

There are published studies indicating that a real display of the northern lights is followed by a southern shift in winds and generation of storms. It may be that storms follow aurora displays in England 2 days later because of the generation of English storms in the "Icelandic low." In Scotland it is a common saying that "The first great aurora of the fall is followed by a storm in two days."

Sir William Herschel noticed that since auroras made the stars twinkle, it was disturbing the upper atmosphere. Many arctic explorers have noted that after a real display of the northern lights, the sky was covered by a thin veil of cirrostratus clouds. These clouds then deepened into a real storm.

It has recently been found by scientists that a major aurora deepens depressions in the Gulf of Alaska and enlarges the storm systems that later strike the United States and Canada. It takes about 10 days for this storm to reach the middle of the United States. That is probably why it has been believed for so long that the aurora had no connection with the weather.

Since the sun rotates on its axis in a 27-day period, a sunspot producing an active aurora may form another storm 27 days later. This gives the weather some sort of periodicity, until that magnetic spot looses its strength. The heat that the sun gives to the earth remains relatively constant, and the increased particles that drive the aurora should not affect the heat balance at all. It is a real mystery why an absence of sunspots generally mean a cold winter, whereas active sunspots mean a period of cold weather.

EARTHQUAKE WEATHER

The three suns were almost in a straight line apparently several toises distance from each other, the real one in the middle, and the others on each side. All three were crowned by a rainbow, the colors of which were not definitely fixed, it now appeared iris hued and now of a luminous white as if an exceedingly strong light had been at a short distance underneath.

. . . Beside the roaring which constantly preceded and accompanied the earthquake we saw spectors and fiery phantoms bearing torches in their hands. Pikes and lances of fire were seen waving in the air, and burning brands darting down to our houses.

—SOME OF THE SIGNS OF THE GREAT CANADIAN EARTHQUAKE OF FEBRUARY 5, 1663,
AS RECORDED BY THE JESUIT FATHERS.

Can you look up in the sky and say, "An earthquake is coming"? It seems improbable, but there is a connection between the weather and earthquakes. The oldest observation of "earthquake weather" comes from the "Cannibal Hymn" in the Egyptian pyramid texts. It says "The bones of Akeru [the sky and earthquake god] trembled with the clouding of the sky after Akeru was transformed into a god."

Aristotle wrote, "Earthquakes are sometimes preceded during the day or after sundown in clear weather by a thin cloud layer that spreads out into space. Besides the weakening of the sun, and the darkness that comes without clouds, the calmness and cold that occasionally occur before earthquakes which happen in the morning confirm the cause."

Pliny collected thousands of nature observations for his books. On earthquakes he wrote, "There is also a sign in the sky. When an earthquake is pending, either in daytime or after sunset in fair weather, it is preceded by a thin streak of cloud stretching over a wide space."

The apostles wrote that the crucifixion of Christ was accompanied by an earthquake, during which the sun turned to blood. There are many accounts that mention the reddening of the sun before earthquakes.

The great geographer, Alexander von Humboldt, lived in Venezuela in 1799. He noticed a red fog on the horizon for several successive days. A fog is created by an increase in relative humidity, but in this fog the relative humidity was only 83 percent. During this time, the stars began flickering, and a 12° halo formed around the moon. These unusual displays ended with a powerful earthquake, and three days later the reddish fogs disappeared.

Humboldt wrote in his "Personal Narrative;"

The inhabitants are most firmly convinced of some connection between the state of the atmosphere and the trembling of the ground. I was much struck by this when mentioning to some people at Copaipo, Chile, that there had been a sharp shock at Coquimbo. They cried out, "How fortunate. There will be plenty of pasture this year." To their minds an earthquake foretold rain, as surely as rain foretold abundant pasture.

Similar observations have been noted in Europe. Immanuel Kant wrote that the great Portuguese earthquake of 1755 was preceded by a red fog and followed by a red rain. That might have been due to increased microbial activity, which has been true on other occasions.

The old Roman and Greek sailors kept a frequent watch on the Stromboli volcano of Sicily. It is called the "Lighthouse of the Mediterranean," because of its frequent eruptions. Aeolus, the God of the Winds, was said to have learned to predict wind directions 3 days in advance by watching its smoke.

There was an old belief that abnormal volcanic smoke meant an earthquake was imminent. Volcanic activity reflects strain in the earth's crust. The gasses would condense into fog more readily before earthquakes and be visible at greater distances.

California skies were their usual shade of blue on the afternoon of March 10, 1933. Streaks of clouds quickly formed over the sky, and within

half an hour, it was completely overcast. No winds were present, and many people noticed this unusual clouding. An hour later, a strong earthquake hit the area.

The San Francisco earthquake of 1906 gave few warnings to residents. The one sign that several people noticed was the abrupt stop in the usual evening land breeze. Before earthquakes in South America occur, the sea breeze stops, too. This indicates that ions must change the density or temperature of the air over land.

In Japanese proverbial lore, earthquakes are followed by wind and rain. The weather that follows a series of earthquakes was studied in 1920. On any given day, the rain probability is 46 percent. For the two days following the quake it rises to 47 percent and then 53 percent. The probability of winds above 13 miles per hour is 25 percent, but on the second day afterward this rises to 63 percent.

Another Japanese proverb reads, "Before an earthquake, the *chiki* [earth air] comes out of the ground." Miners notice the *chiki* in the mines and stop work. Other well-known Japanese earthquake signs are an extended sun, which is distorted and reddish. The stars glow like smoking flames, and the sea breeze ceases. There are thin wisps of clouds like jet-vapor trails that form above fault lines, and dense, ground-fogs that form over fault lines as well.

It is of great interest to find that earthquakes seem to follow long-range weather patterns in Japan. When there are years of maximum rain and

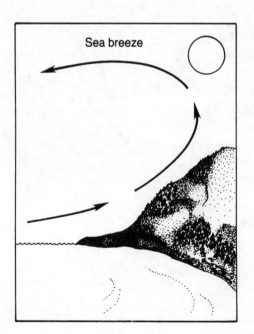

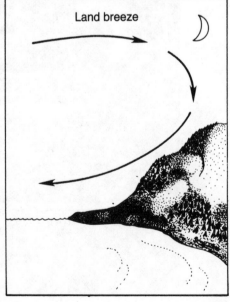

snow, many earthquakes occur. When dry weather prevails, there are relatively few earthquakes. This appears to indicate that weather and earthquakes are driven by outside forces.

In 1895, a Cambridge University student published a paper that may have indicated the source of earthquake-weather phenomena. He was breaking down water into hydrogen and oxygen by electrolysis. The uncombined gasses leaving the electrodes were condensing atmospheric moisture into fog at a relative humidity as low as 80 percent.

Before earthquakes, rocks are squeezed along fault lines. This generates electricity. The current breaks down trapped water and releases large amounts of hydrogen and oxygen ions. These ions condense water vapor, and this causes fog and clouds. The winds and storms that follow may be a result of ions acting like a Faraday motor.

The best way to predict earthquakes may not be through seismographs, but through earth satellites. The blurry white or reddish glow noticed by so many observers ought to be detectable by using special spectrum filters on satellites. The sky may contain omens of the future, and it may give us fair warning of the earth's intentions.

THE ELECTRICAL SKY

When we passed Crete only sky and water could be seen, for Zeus brought a dark cloud to rest above the ship. The sea darkened below it. Zeus thundered and in the same moment struck the vessel with lightning. The ship reeled to the blow of his bolt and was filled with sulphur.

Then Poseidon drove together the clouds and stirred up the sea. In his hands the Trident; he excited violent squalls, among all the winds. With dense clouds he covered earth and sea; and night advanced from the sky. Together broke the east wind, the south wind and the stormy west wind. And the north winds from cold lands came, driving huge waves.

—FROM HOMER'S *ODYSSEY.*

It was Benjamin Franklin who established that sparks caused by stroking a cat on a dry day and lightning from the sky are the same thing. It wasn't until the twentieth century that we began to understand that electrons were the negative building blocks of matter and protons were the positive building blocks.

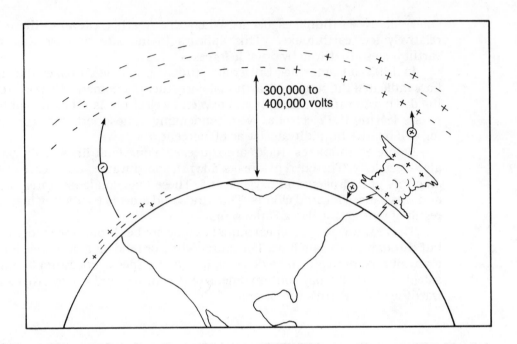

The Greeks thought that it was Zeus who hurled thunderbolts from the sky to punish his enemies. The symbol of Zeus can be seen on the dollar bill: an eagle with a lightning bolt in his claws. Without any understanding of electricity, primitive man could only conclude that lightning belonged to God.

Early scientists found that the electrical charge of one's feet is 200 volts more negative than one's head on a normal day. This presented a great paradox, because it became obvious that electricity was always leaking into the air, yet the ground remained negative. We don't get electrocuted from the 200-volt charge differential because there is almost no current passing through our bodies.

The mystery of the electrically conductive sky was solved in 1911 when Victor Hess, a Viennese physicist, sent a balloon into the upper atmosphere. He found that cosmic rays were ionizing the earth's atmosphere and making it electrically conductive. He won the Nobel Prize for this observation in 1936. The discovery of radioactive elements around the same time, and the fact that these elements turn air from a nonconductor into a conductor, contributed to the mystery of the leaking current.

When scientists were able to measure the charges on raindrops, they found that rain in different areas of storms was differently charged. Negatively charged rain had a slow, uniform rate of fall. The heaviest rainfall, in the center of the storms, was generally positively charged. It is believed that

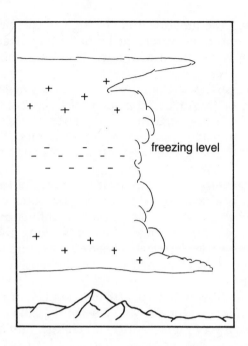

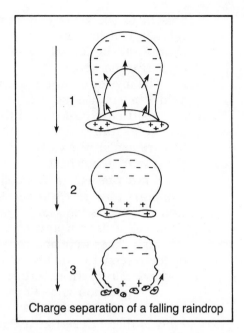

Charge separation of a falling raindrop

water vapor condenses more readily around negative ions than positive ions. At the edges of the storm, negative ions aid in coalescing water vapor.

The electricity of storms proved to be one of the most difficult scientific problems to understand. A large cumulonimbus cloud has a pancake-shaped layer of negative charge that may be a mile thick and 5 miles in diameter. Above this layer, the entire upper part of the cloud is positively charged. The separation of the charges accounts for lightning bolts. The mystery was in finding out how the charges separate.

There are two explanations for charge separation. Raindrops are remarkably uniform in size. If a drop is slightly oversized, the bottom part of that drop shears away in a spray of fine droplets. The negative charge on the drop orients toward the positive sky; the bottom area of the drop is positive. The oversized drops at the edges of the storm thus eliminate their positive charge and fall with a negative charge.

The main mechanism of charge separation seems to be the freezing of moisture. The heat of cumulus clouds carries air upward. As the air rises, it expands, cools, and eventually freezes the water it carries. Below the freezing level, the cloud is charged negatively; above this level, it is positively charged. The regular crystalline structure formed by freezing squeezes out the electrons. Falling snow is almost always positively charged.

When ice was frozen out of water in the laboratory, physicists found that the ice was 70 volts positive in relationship to the water. The salts present in the water determined the amount and the sign of the electrical charge. If ammonia was present in the water, the ice had a high negative charge. If table salt was present in the water, the ice took on a positive charge.

The mystery of the negative earth was solved when scientists began to study the charges on lightning bolts. About 16 percent of lightning bolts are positively charged, and the rest are negative. At any given time, there are 3,600 thunderstorms on our planet producing negative bolts. These bolts replenish the earth's negative field and keep our feet 200 volts more negative than our heads.

It must be asked, "Does electricity bring rain, or it is a result of rain?" It appears that the electrical sky is a result of rain and freezing water in the clouds. Naval scientists at Jacksboro, Texas, attempted to influence the electrical fields of clouds by using a high-voltage wire 4 miles long and 30 feet high. They were able to reverse the electrical fields of clouds, and some of the reversed clouds were said to have turned into thundershowers far downwind from the experiment.

The question is still open as to whether electrifying clouds can bring rain. From our study of earthquake weather, it would seem that the possibility is there. Desert areas have a very high positive-ion charge, and areas of the Northwest with heavy rains have negative charges. Electrical rain making could be one of the "miracles" of the future.

THE NEON SKY

Last night I saw St. Elmo's stars,
With their glittering lanterns all at play.
On the tops of masts and the tips of spars,
And I knew we should have foul weather today.
> —HENRY W. LONGFELLOW, "GOLDEN LEGEND"

The light thou beholdest, streams through the Heaven.
In flashes of crimson, is but my red beard.
Blown by the night wind, afrighting the nations.
> —HENRY W. LONGFELLOW, "THE CHALLENGE OF THOR"

In the month of February about the second watch of the night, there suddenly arose a thick cloud followed by a shower of hail; and the same night the points of the spears belonging to the fifth legion seemed to take fire.
> —"THE COMMENTARIES OF JULIUS CAESAR"

Peter Kalm was the first great naturalist to visit America. He was a personal friend of Linnaeus, and his main mission was to search for new plants and medicines. He sent thousands of specimens to Europe, and Swedish magazines published his accounts of buffalo, Indians, and rattlesnakes.

When he visited "New Sweden," he found that the settlers made a clear distinction between the northern lights and the "sno-eld" or "snow fire." If the sky was overcast at night, a reddish light could be seen near the horizon, which resembled a distant burning building. In winter, the settlers considered it to be a sure sign of snow; in summer, it indicated rain. The snow fire almost always occurs in the southwest direction.

Kalm observed a number of snow fires and concluded that they were a good sign of snow, although sometimes the snow fell at some distance away. The early settlers originally believed that these snow fires were caused by Indians, until they learned that it was a weather sign. I myself have seen faint snow fires across the prairies of Iowa. They can hardly be recognized anymore because of the presence of artificial light.

In other countries, people spoke of the "glower" or "storm lights" that presaged storms. All of these light phenomena are similar to the glow produced in a fluorescent bulb when electrons strike gasses and give off light.

The most familiar phenomenon of the glowing sky is called St. Elmo's fire. The ancient Greeks knew of this phenomenon and called a single jet "Helena" and a double glowing flame "Castor and Pollux." Helena was the sister of Zeus, and Castor and Pollux were the sons of Leda. The city of Sparta dedicated two gold stars to the shrine at Delphi to honor Castor and Pollux.

Why did this mysterious glow become associated with Saint Elmo and become a sign of good fortune? The story begins with an Italian bishop who was killed for his Christian beliefs in A.D. 304. Ten centuries later, Saint Eracemus, a Dominican monk, adopted the bishop as his patron saint. After he died, sailors adopted the bishop's name in place of Eracemus and begin calling the mystery fire Saint Elmo.

Christopher Columbus wrote of this fire on his second voyage: "On Saturday at night, the body of St. Elmo was seen, with seven lighted candles in the round top, and there followed mighty rain and frightful thunder. I mean the lights were seen which the seamen affirm to be the body of St. Elmo, and they sang litanies and prayers to him looking upon it as most certain that in these storms when he appears, there can be no danger."

When Fernando Magellan made his around-the-world trip, his recorder noted: "During those storms the holy body, that is to say St. Elmo, appeared to us many times in light . . . on an exceedingly dark night on the maintop where he stayed for about two hours or more to our consolation, for we were weeping. When that blessed light was about to leave us . . . we called for mercy. And when we thought we were dead men, the sea suddenly grew calm."

This ghostly light has turned large ships into glowing light. It is not usually associated with lightning, which seems strange, for both are an electrical discharge. There are hardly any photographs of it, for although the glow is visible to the naked eye, the soft light is difficult to capture on film.

There is an old account of a traveler on a stagecoach heading toward a remote Idaho mining town. Each time the driver raised his whip to spur the reluctant mules, the tip would glow. His passenger found that when he raised the tip above 8 feet, it would begin to glow. The glow is based on electrical potential, and the higher an object is extended, the greater the electrical potential is in respect to the ground. This is the reason why the old sailors were so familiar with St. Elmo. Their high sailing masts created a large potential difference and, without any lights, they readily noticed St. Elmo. In the last part of the storm, the ship would pick up a positive charge and began discharging into the low, negatively charged clouds.

Records were kept of the occurrence of this glow at the weather station of Ben Nevis in northern Scotland. They found that the glow normally occurred 6 hours after the center of a storm passed. Its presence was almost always associated with showers of graupel. It could not be seen during the day, but on several occasions, its presence was known because of the hissing at the top of the weather station.

These observations explain why the mysterious light was always associated with good luck. On all of twenty occasions, with one exception, the center of the storm had passed and better weather was coming.

There are a number of observations of entire clouds glowing with such intensity that papers could be read by their light at night. There may be a simple explanation for this glow. The storm separates positive and negative droplets. When positive and negative droplets are brought together in the laboratory, there is an electrical discharge, and light is emitted. It is possible that turbulence in the clouds at night unites the charged particles and produces the neon sky.

RADIO WEATHER

Fourth Movement from Beethoven's *Pastoral Symphony*

Grand Canyon Suite, by Ferde Grofe

Scheherezade, by Nikolai Rimsky-Korsakov
LISTENING TO MUSICAL STORMS ON YOUR STEREO

You can listen to the weatherman on the radio, or use the radio to determine future weather. Amplitude-modulated (AM) radios have a circuit that picks up storm-generated static. Frequency-modulated (FM) radios are designed to filter out radio static.

By 1895, the first radio pioneers had found that the clicks in their primitive receivers coincided with lightning flashes. This led to the idea that weather forecasts could be made using radio receivers, and a serious attempt was made in England after World War II to set up a radio network to improve weather forecasting.

The early radio pioneers called their unwanted radio noise "X's," "clicks," "hisses," "cracks," and "grinders." They became known as *atmospherics*, and this was shortened to *sferics*. These *early radio* pioneers believed they could tell the nature of future weather from the radio noise on the shortwave bands. Weak cracklings preceded frosts or temperature falls. Cold-front thunderstorms signaled their approach with violent cracklings. Hail storms produced a slight hiss.

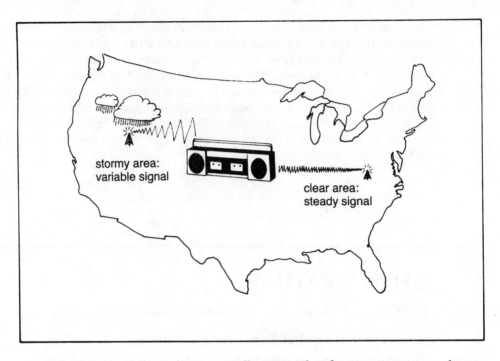

stormy area:
variable signal

clear area:
steady signal

A cheap transistor radio, especially one with a ferrite antenna, makes a good detector of sferics. This kind of antenna is highly directional, so the source of static can be located by turning the radio. Tune the radio to an empty spot on the low end of the band and turn up the volume. Faint cracklings are a sign of good weather; definite crackles mean that electrical activity is several hours away. As the storm approaches, the crackling increases. The high side of the AM band produces less crackling than the low side, but when the storm gets nearer, the amount almost equals that of the low side.

A circuit that detects and plots sferics was written up in the March, 1959, issue of *Scientific American*. When measurements of electrical activity in the sky were combined with wind and barometer readings, nearly every forecast was accurate. In the Midwest, banks of cumulus clouds build up on many afternoons. They usually dissipate with the setting sun, but they can turn into good thunderstorms. The sferic detector could tell if they would turn into storms at a time when the weather bureau was only 50% accurate on its afternoon forecast.

The electrically charged layer above our atmosphere is known as the *ionosphere*. During daytime, solar radiation excites particles, and the ionosphere drops. At night, it rises and becomes a reflector of radio signals. It is for this reason that distant radio stations begin to crowd into the radio band after the sun sets, while many small, local stations sign off for the night.

The early ham-radio operators found that the state of the sky made a big difference in how far they could receive and transmit. Several studies produced the following results:

Sender and receiver in cloudy area: good transmission 83% of the time
Sender cloudy and receiver clear: good transmission 83% of the time
Sender and receiver in clear weather: good transmission 46% of the time
Sender clear and receiver cloudy: good transmission 46% of the time

Physicists in West Virginia found that a nearby radio station could be monitored and used to give accurate weather forecasts. Since the signal bounced off the ionosphere, this meant the electrical layer must be influenced by or be influencing, the changing atmospheric conditions.

The signal intensity of the Morgantown, Virginia, radio station was plotted against the coming weather. An increase in signal strength meant cloudy weather, and a decrease indicated better weather. If the night-signal strength rose for several hours and then fell, it meant that the morning would be cloudy, and the afternoon would be clearing. During February 1929, the forecasts were 90 percent accurate and only one was wrong. During March, they were 88 percent accurate, with 3 wrong forecasts. This method provided forecasts for future weather on a short term basis that were more accurate than those of the normal announcer.

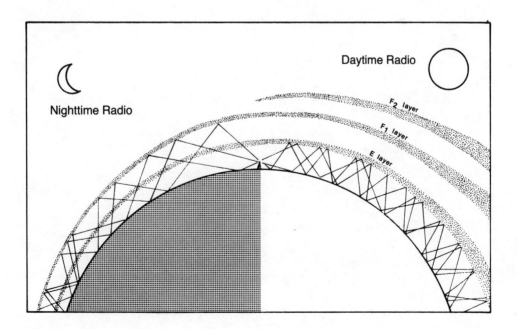

Another highly provocative bit of radio detective work was done by Father Ernest Gherzi in China. He sent a 6 Megahertz signal from an ordinary radio transmitter just after sunset and just before sunrise. The length of time it took for the signal to return indicated which ionosphere layer had bounced it back. A signal returning from the E layer meant maritime air with overcast and rainy weather. An F-1 layer return meant cold, dry Siberian air was over the area. A return from the F-2 layer meant that a warm, tropical air mass was present.

Radio signal weather forecasting has been neglected, but it was used during World War II. There was a 7-mile-long plant at La Porte, Indiana, which produced shells and bombs. Because of the danger of explosions, the plant was shut down during thunderstorms. Radio static was monitored every half-hour, and as electrical activity increased, the plant was shut down in sections.

The sound of the skies has been neglected because so many other areas have been studied. If a thorough study were to be made of the unusual radio sounds from hail, frost, and other phenomena, we might be able to produce more accurate predictions and to acquire a better understanding of the voice of nature.

IV.
STORM
WARNINGS

THERMALS AND THUNDERSTORMS

The Rainy Day

My life is cold, and dark, and dreary;
It rains and the wind is never weary;
My thoughts still cling to the mouldering Past,
But the hopes of youth fall thick in the blast,
   And the days are dark and dreary.

Be still, sad heart! and cease repining;
behind the clouds is the sun still shining;
Thy fate is the common fate of all,
Into each life some rain must fall,
   Some days must be dark and dreary.

—HENRY WADSWORTH LONGFELLOW

The beginnings of the thunderstorm are invisible to human eyes, although birds can sense them. These are the *thermals*, rising currents of warm air coming from heated land. These masses of hot air cool 1°F for every 300 feet that they rise. At some point in the sky, the water vapor in the thermals reaches the dew point. The air continues to ascend for another 250 feet until it suddenly forms a billowing cloud.

The pyramids of Egypt generate clouds from the hot air ascending along their slopes. Mountains, hills, and large parking lots are natural generators of ascending hot air. The steel mills of Indiana release so much heat that rainfall increases by about 20 percent miles downwind from the plants.

It takes much energy for a bird to flap for long periods of time, so nearly all prey-seeking birds ride the thermals. Vultures and condors wait until the first good gusts of wind begin in the morning before taking flight. They flap around in big circles as they gain altitude. When they have found the rising air, they begin flying in trochoidal circles without moving their wings. Can birds see thermals? According to glider pilots, gliding birds are able to head directly for distant areas of uprising air.

William Espy was the first person to recognize the nature of hot air in the generation of thunderstorms. He collected accounts of storms that had been generated by forest and prairie fires. He felt that the Great Plains could be made fertile by burning 40 acres of timber at various intervals when rain was needed.

Local thunderstorms feed on the heat of rising air. It takes the same amount of heat to turn 1 pound of boiling water to steam as it takes to raise

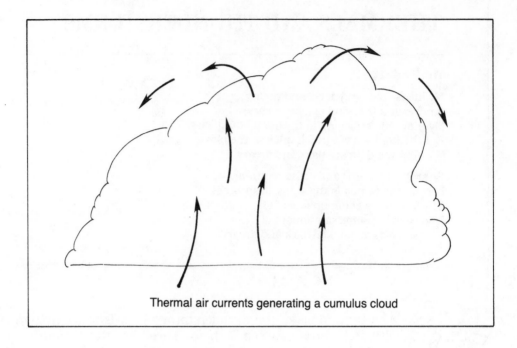

Thermal air currents generating a cumulus cloud

5.36 pounds of water from freezing to boiling. When vapor turns back into water, heat is released, so storms are really heat-energy releasers. The rising center of the storm is warmer. This causes updrafts, which condense water and make rain.

Espy wanted to take his ideas of storm generation and forecasting and create a national weather service. He contacted influential congressmen and met with the former President John Quincy Adams. Adams wrote in his memoir: "Mr. Espy, the storm breeder . . . The man is methodically monomaniac and the dimensions of his organ of self esteem have been swollen to the size of a goiter by a report of a committee of the National Institute of France, endorsing all of his crackbrained discoveries in meteorology."

Each method of storm generation depends upon the generation of rising air, which then cools and condenses the moisture within it. When air flows over mountains, it rises, cools, and dumps its moisture. When it flows down the other side of the mountains, it warms as it goes lower; this side is dry.

The air mass theory of Bjerkness is based on the uplifting of air when air masses of different temperature meet. Warm air is lighter than cold air. When it is pushed against a cold mass of air, it flows over the top, cools, and releases a gentle rain.

When cold air approaches a stationary mass of warm air, the cold, dense air pushes the light, warm air upward. This produces heavy rains and violent electrical activity. Cold fronts in the United States and Canada often have high winds and tornadoes. These do not generally occur in Europe and Asia.

Heat thunderstorms are not generated by air masses. These storm *cells* spread a trail of rain for perhaps 5 miles across and maybe a hundred miles long before they die in the late afternoon. The next day, that strip will be cool, but the adjacent strip will generate a thunderstorm and suck in damp air from the sides. This means that scattered thundershowers are likely to hit the dry areas on following days.

When the forecaster says, "There is a 50 percent chance of scattered showers," he is making a *probability forecast.* Since the showers are isolated cells, they will not cover the area as a front does, but will scatter themselves through half of the area.

Simple thunderstorm prediction is based on many factors, but with only a wet-bulb thermometer and a barometer, it is possible to predict accurately the chances of having a thunderstorm in the afternoon. A combination of a high wet-bulb temperature and a drop in the barometer raises the probability of an afternoon thundershower.

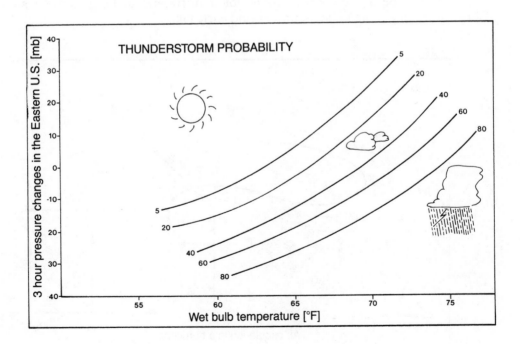

THE MIGHTY HURRICANE

June—Too soon.
July—Stand by.
August—You must.
September—Remember.
October—All over.

—This is the little poem that all good Caribbean
sailors remember during the hurricane season.

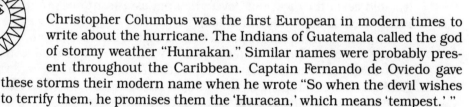

Christopher Columbus was the first European in modern times to write about the hurricane. The Indians of Guatemala called the god of stormy weather "Hunrakan." Similar names were probably present throughout the Caribbean. Captain Fernando de Oviedo gave these storms their modern name when he wrote "So when the devil wishes to terrify them, he promises them the 'Huracan,' which means 'tempest.' "

The same storms in other parts of the world are known as *typhoons, baquiros, Bengal cyclones* and *willy-willies.* The ocean-water temperature has to be above 79°F in order for a hurricane to be generated so they normally come in late summer and early fall.

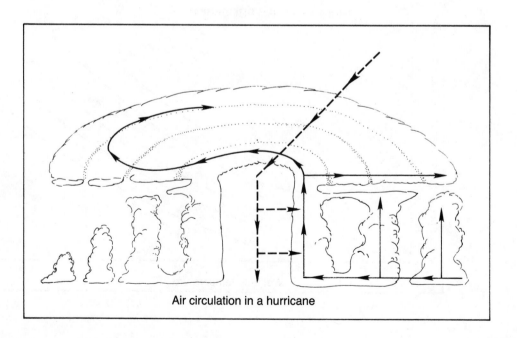

Air circulation in a hurricane

Meteorologists use the term *tropical storm* when a storm's winds are under 70 miles per hour, and *hurricane* when the wind speed rises. Hurricanes are marked by an *eye,* which stretches from 10 to 30 miles wide and contains warm temperatures and clear skies. Around this tropical paradise rages an inferno of winds gusting at speeds up to 186 miles per hour. If 1 percent of the energy in one hurricane could be captured, all of the power, fuel, and heating requirements of the United States could be met for an entire year. It takes 500 trillion horsepower to whirl the great core of winds at such tremendous speeds. It is the equivalent of exploding an atomic bomb every 10 seconds.

The classical sign of these storms are blood-red and green streaks on the distant horizon. An old English captain wrote: "At sunset we had a beautiful sky to westward, light hazy, cloudy shaded from deep crimson to the lightest pink with streaks of green between them. Near the horizon the green was of a very deep colour. My passengers were all admiring it. I told them that the old sailors said that green in the sky betokened no good and so it proved with us."

At night the stars have a peculiar dancing appearance. Stars that are red or blue will suddenly become silvery and whitish and then reddish or blue. The moon is usually surrounded by intense halos.

Sooty Tern

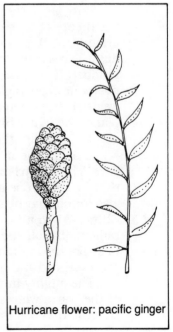

Hurricane flower: pacific ginger

The surface of the sea becomes covered with scum and weeds. The sea smells unpleasant at this time. The low pressure and intense wave action stir up the bottom and bring seaweed and organic matter to the surface.

At high altitudes, the storm projects a tongue in the direction it is traveling. This area of low pressure is marked by crisscross puffs of cirrus clouds. At ground levels, the winds are constantly shifting, and a red haze covers the sun.

The Pacific islands have few time markers to indicate seasonal weather, but they do have the hurricane flower. It is a close relative of ginger and is used for food. The flowers are normally small and white, but during hurricane season, they turn crimson.

Hurricanes in the Caribbean are marked by the arrival of millions of weather-wise birds. They are sooty terns, also known as "hurricane birds." In the Dominican Republic they are called "twa-oo" after their cries. They leave the low, sandy islands where they breed and cover the rocky shores hundreds of miles away. Ships in the area of hurricanes are sometimes covered with these birds.

Why should a tropical storm grow into a frenzied sea monster? And why do many storms dissipate before they reach that stage? One interesting conjecture is that solar flares increase high-altitude winds, which feed the storms. The Russian meteorologist Anatoli Djakov is said to have predicted many storms successfully, days in advance. He monitors sunspot activity closely, particularly the spots near the central meridian of the sun, for their particles are on a direct line with the earth.

Karl Mierback, an Austrian astronomer, was among the first to hold the theory of flare-generated tropical storms. He invoked this theory in an attempt to stop two French pilots who planned to make the first nonstop flight across the Atlantic in May 1927. They were hoping to beat Charles Lindbergh. Mierback sent a telegram to the French pilots asking them to postpone their flight because of large solar flares. The head of the French weather service read the telegram and replied "Nonsense." The French plane, the *White Bird,* disappeared without a trace when unexpected tropical storms swept the Atlantic.

A hurricane may hover near coastlines, keeping residents worried for days. Then it strikes the land with an iron blow. Ernest Gherzi, who found that he could use a simple radio transmitter to predict weather patterns, also found that he could predict whether hurricanes would hit the coast of China. When a hurricane was still 200 miles away, he bounced a 20-watt radio signal off the ionosphere. If it hit the E layer, a hurricane was likely to strike the mainland. If the radio signal bounced off the F layer, polar air was overhead and the storm would return to sea.

The mighty hurricanes are energy parasites. They turn the heat energy of the warm waters of the Caribbean into screaming winds. Experiments suggest that we can seed these storms and remove the water that gives them such titanic amounts of energy. But the acts of man still have a long way to go to tame an "act of God."

STRANGE RAINS

"It rained fire and brimstone from heaven."—Luke 17:29
"It rain'd downe fortune showring on your head."—William Shakespeare, A.D. 1596
"It shall rain dogs and polecats.."—Richard Brome, A.D. 1653
"He were sure it would rain cats and dogs."—Jonathan Swift, A.D. 1738
"It's raining pitchforks and hammerhandles."—Jake Falstaff, A.D. 1938
—RAIN EXPRESSIONS IN LITERATURE.

During a downpour we say that it rains "cats and dogs." There are several theories about this rainfall saying. It is possible that the word *cats* is derived from the Greek word *catadupe* meaning "waterfall." Or it could be raining *cata doxas*, which is Latin for "contrary to experience," or an unusual fall of rain.

In 1870, Brewer's *Dictionary of Phrase and Fable* published an interesting explanation for our favorite rain expression:

> In Northern mythology the cat is supposed to have great influence on the weather, and English sailors still say, "the cat has a gale of wind in her tail," when she is unusually frisky. Witches that rode upon the storms were said to assume the form of cats; and the stormy northwest wind is called the "cat's nose" in the Harz mountains even at the present day.
>
> The dog is a signal of wind, like the wolf, both which animals were attendants of Odin the storm-god. In old German pictures the wind is figured as the "head of a dog or wolf," from which blasts issue.
>
> The cat therefore symbolizes the down-pouring of rain, and the dog the strong gusts of wind which accompany a rain-storm; and a "rain of cats and dogs" is a heavy rain with wind.

Ancient history is filled with accounts of strange things falling from the sky. Pliny the Elder described a number of these showers at the time of Christ in his *Natural History*. He describes rain of wool, flesh, and birds, all of which were looked upon as supernatural omens. Rains of milk and blood are more likely due to dust from the Sahara desert or a volcanic eruption.

The Bible describes a rain of manna and quails more than 3,000 years ago. Although this was looked upon as a supernatural event, it was actually not an uncommon thing. The rain of manna has happened frequently in modern times; the manna is really a lichen that grows in great numbers after rains.

A "rain of blood" is also described in the Bible and has happened many times throughout history. Some of the red rains are iron dust picked up by desert whirlwinds. There is also a red algae, which possibly multiplies in the sky or on the ground just after it rains. In northern regions, the top layer of snow is often red due to red snow algae.

There are a number of accounts of yellow or sulphur rains. In each instance wherein someone studied the yellow color, it proved to be pollen. Apparently the wind conditions were light when the pollen was ripening. Then a heavy wind stripped the pine forests and deposited it elsewhere as "yellow rain."

There are numerous accounts of rains of frogs, hay, fish, and grain. All of these accounts seem to be due to whirlwinds. A good whirlwind can lift thousands of pounds and carry objects for miles. There is one reliable account of a fishing boat that sailed into a large waterspout. Fish flew everywhere.

There are about 70 recorded rains of fish, but nearly all of the rains of fish are small ones. There is, however, one account of a fish fall in India in which more than 10 people picked up fish weighing up to 8 pounds each.

In Tahiti, the natives used tanks on their houses to collect rain water, and they would often find small fish swimming about. They called the "rain-fish" *topataua,* which means "rain-dropped."

In 1939, the *Times* of London published an interesting letter. The writer referred to several showers of fish that had fallen on the English countryside and suggested that whirlwinds hadn't really carried these fish along with the storm. Instead, they had sucked up fish eggs. The eggs had then germinated in the clouds. Finally, when the developing fish got too heavy, they fell to earth.

Alexander von Humboldt investigated several accounts of fish erupting from volcanoes in South America. The events happened wherever there were crater lakes. During the eruption, fish and mud were spewed about for miles.

There are many accounts of rains of ice-coated ducks, grasshoppers, and fish, but there is no account of a rain of cats and dogs. But there is an old English joke that asks, "What's worse than raining cats and dogs?" The answer is "hailing taxicabs."

LIGHTNING AND THUNDER

If the first thunder is in the south, Aha! The bear has stretched his right leg in the winter bed.

If the first thunder is in the west, Aha! The bear has stretched his left arm in his winter bed.

If the first thunder is in the north, Aha! The bear has stretched his left leg in his winter bed.

If the first thunder is in the east, Aha! The bear has stretched his right arm and comes forth and the winter is over.

—ZUNI INDIAN PROVERBS

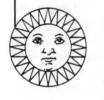

In earlier times, nearly all people believed that thunder and lightning were the play of the gods. In Scandinavia, the great red-bearded Thor was swinging his hammer. When Thor swung his hammer, the mixture of water and *vafermest* (mist-fire), as the Vikings called it, separated, and a stroke of lightning split the skies.

The Greeks believed that Zeus was playing with the symbols of his power. The Roman poet Lucretius was one of the first great rational skeptics. Why did Zeus have to wait for thick clouds before displaying his power? Why did he scatter his bolts on mountains, seas, and even the temples belonging to him if he controlled them?

The belief in the old gods died hard, for even until recent times the skies were controlled by Jehovah or Jesus. When Constantine converted to Christianity, he made a law authorizing the Romans to continue consulting their oracles when lightning struck a building.

If you are close to a lightning strike, you might hear a *clik* or a *vit* sound. Other people have described the sound as a swishing or a tearing of cloth. The old sailors who heard this sound thought that the sails were being ripped apart. This is the true sound of lightning, not the thunder as we usually think.

Lightning leaves behind a pungent, sulphurous smell. The smell is due to two chemical reactions: one that turns oxygen into ozone and another that forms nitrous oxides. These compounds furnish a significant amount of organic nitrogen, good for growing crops.

We see lightning as a white streak or an icy blue light before the storm. It is an old observation of forest rangers that white lightning causes forest fires, but red lightning does not. When an electrical stroke travels through

108

rain, it ionizes some rain and produces a reddish color, so this belief is true.

There are a number of observations about the color of lightning, some of which may have some validity. They may depend on the distance and the cloud cover between you and the stroke. Bluish lightning is associated with hail. Storms begin with red lightning and end with yellowish strokes. When lightning is white, there is a single boom of thunder, but red lightning is accompanied by rolling thunder.

When lightning strikes the ground, there is a sharp *crack*; this is most often the case with cold-front storms. Warm-front storms produce strokes that run through the length of the cloud, and the noise rolls. This is the basis of the Arabic proverb, "Sharp thunder means improving weather, and rolling thunder means bad weather." A cold front takes a short time to pass, and good weather returns quickly.

Satellites have discovered "superbolts." A strong, high-altitude wind blows off the top of a cloud and builds an enormous charge. It discharges in a stroke hundreds of times above normal strokes. These are rare, and even severe storms are unlikely to produce more than one.

The air literally explodes with the heat of a bolt, and this energy goes into the thunder. Only a fraction of the energy goes into noise; even then, much of that is below the threshold of human hearing. It takes 10 million trumpeters to produce the equivalent of 1 horsepower in sound. It would

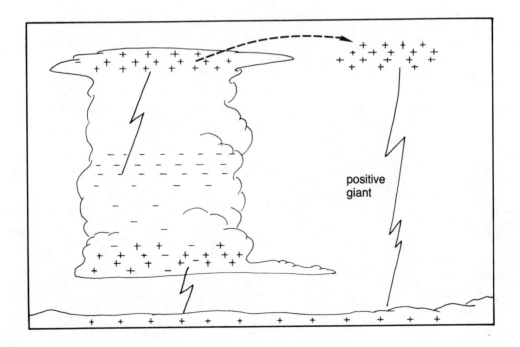

positive giant

take 200 million trumpeters blowing for 13 seconds to produce the same amount of noise as is present in 1 good crack of thunder.

There is an old belief that thunder generates thunderbolts. But there are no real thunderbolts. These objects have been shown to be meteorites or other objects that people thought came with the storm. Lightning does strike sand and fuse it into vase-shaped objects known as *fulgurites*. These have been found often in desert areas where rainstorms wash away sand and expose them.

Animals can often sense lightning a second or so before it happens. Chickens scatter and dogs bark before a bolt strikes nearby. There is an account of fireflies blinking a second or so before a flash. They are probably sensing the electrical build-up before the stroke. It is the same phenomenon as your hair standing on end in response to electricity.

GREEN LIGHTNING RODS

. . . Onditachiae is renowned among the (Hurons) like a Jupiter among the heathen of former times having in hand the rains, the winds, and the thunder. This thunder is by his account, a man like a Turkey; the sky is his palace, and he retires there when it is serene. He comes to earth to get his supply of adders and serpents, and of all they call Oki. When the clouds are rumbling, the lightnings occur in proportion as he extends his wings. If the uproar is a little louder, it is his little ones who accompany him, and help him to make a noise as best they can.

Raising an objection to him who told me the story, "where comes dryness?" "It comes from the caterpillars, over which Onditachiae has no power." "Why does the lightning strike trees?" "It is there, that it lays in its supply." "Why does it burn cabins, why does it kill men?" "How do I know?"

—THE WRITINGS OF THE EARLY JESUIT PRIESTS WERE COLLECTED INTO
A SERIES OF BOOKS CALLED *THE JESUIT RELATIONS*.
THIS IS LE JEUNE'S RELATION, WRITTEN IN 1636 IN CANADA.

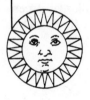

Are there any safe trees to stand under during a storm? Most of us have been taught as children not to stand under any trees during a storm. The tallest object in the area is believed to be the most likely to be struck. Or is this simply superstition?

The Roman emperors earned their laurels and, on occasion, wore crowns made of laurel branches. The laurel was sacred to Apollo, and as a result was believed to be immune to lightning. During thunderstorms, Emperor Tiberius was afraid of lightning, so he wore his laurel crown to be on the safe side.

The tree that was most sacred to Zeus was the oak. At Dodona, Zeus gave predictions by the rustling of oak leaves. The emblem of Zeus on the dollar is a lightning bolt borne by the eagle. In one claw, the eagle holds the bolt; in the other he holds an olive branch, a plant said to be immune from lightning.

The tough Viking warriors respected Thor, who hurled his mighty hammer during storms. Occasionally, parts of his hammer (meteorites) were found when farmers plowed their fields. Thor had three special "thunder plants." These were the thunderbeard, *Sempervivum tectorum*; the thunder vine, *Hedera helix,* and the thunder plant, *Sedum species.* The thunderbeard, also called the houseleak, was planted on the roofs of houses to protect them against lightning.

English folklore described a number of trees and the relative danger of being struck while beneath them:

Beware of the oak, it draws the stroke.

Avoid an ash, it courts the flash.

Creep under the thorn (hawthorn), it will save you from harm.

The truth of these proverbs has been the subject of a number of studies. Below are some records of the number of trees struck in a limited area during a particular year.

America, 1898: oak, 48; pine, 33; spruce, 5; beech 1
Germany, 1920: oak, 56; ash, 20; pine, 4; beech, 0
England, 1935: oak, 61; elm, 32; ash, 26; poplar, 13

A study of the German forests around the year 1900 confirmed the truth of the old proverbs. Given an equal number of trees, 1 beech would be hit to 6 spruce, 37 pines, and 60 oaks. German scientists found that the beech was a poor conductor, but the oak was a good conductor.

The scientists found that a series of factors were involved. If the tree was tall, in the open, bordering the woods, or growing along a stream, its likelihood of being struck increased. Trees growing in sandy soil, and trees having a central taproot extending into groundwater, were more likely to be hit than trees in clay soil with horizontally extended roots.

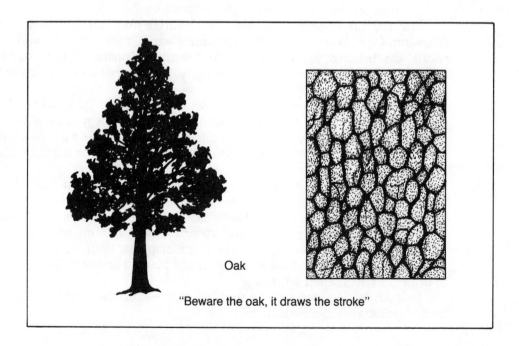

Oak

"Beware the oak, it draws the stroke"

In European forests with an equal mixture of oaks and beeches, it appears that the oaks will be hit 60 times more often. If you don't know the difference between one tree or another, and you feel compelled to stand under a tree in pouring rain, choose a smooth-barked tree. The beech, birch, holly, hornbeam, and maple are all unlikely targets. The rough-barked oak, elm, ash, pine, and chestnut are all likely targets.

A real confirmation of this occurred on May 27, 1964, when a severe thunderstorm struck the town of Tilburg in the Netherlands. A 15-foot-high oak grew 2 feet away from a 70-foot beech tree. Lightning struck the smaller oak, leaving the towering beech tree standing.

The redwoods and the sequoias of California have roughly the same diameter at the base, but the record redwood is 364 feet tall, the record for a sequoia is 272 feet. The redwood is generally taller and more slender, the sequoia shorter and more massive. The big difference in the record heights may be due to thunderstorms. Almost all large sequoias have had their tops shattered by lightning, but in redwood country, storms are rare.

If a tree is hit by lightning before rain begins, it could be shattered by the bolt. If it is hit after rain begins, it might only lose a little bark, for the water provides a conductive surface for the electricity.

There are instances of lightning killing all the brush around a tree within a diameter of up to 18 feet. In one instance, a sycamore tree was

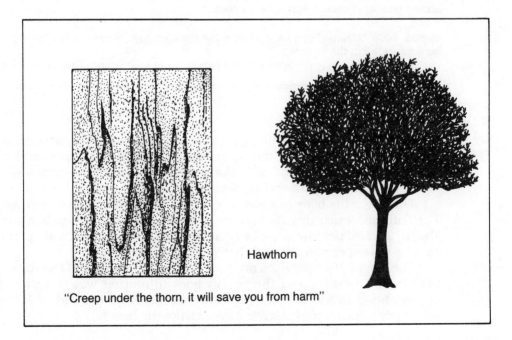

Hawthorn

"Creep under the thorn, it will save you from harm"

struck, and the bark was scored from the top to the bottom of the tree. The tree developed over 200 branches along the lightning line.

Perhaps we could use our knowledge of the electrical effects of the trees for something other than shelter from heavy rains. There have been up to 250 fires started in a single day by thunderstorms in our forests, and it costs millions of dollars to fight these fires. Maybe we could plant lightning-resistant trees and cut down the number of forest fires!

WIND, WATER, AND WEATHER

January 21, 1661: "It's strange we have had this winter no frost at all; the rose bushes are in full leaf; such a time never before was known."

September 19, 1663: "Waked this morning with a very high wind, and said to my wife, 'I pray we hear not of the death of some great person, this wind being so high.' "

January 10, 1666: "News arrived of great storms on the sea. Our fleet scattered; three arrived back at Plymouth with loss of masts."

May 22, 1668: "It rained very hard all this day; the king gone to Newmarket to see the horses race, all in the same wet."

—SOME OF THE WEATHER NOTES CONTAINED IN SAMUEL PEPY'S DIARY.

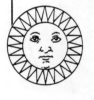

The old-time fishermen were able to predict storms by observing the waves and tides. Storm-generated waves send signals to distant beaches thousands of miles away. Abnormal tides mean stormy disruptions to the ocean basins.

The larger the body of water, the longer its natural wave period will be. The Gulf of Mexico usually has waves spaced about 5 seconds apart, or about 12 waves per minute. The Atlantic coast has a normal wave period of about 8 waves per minute, and the Pacific has about 7.

The height of a wave does not go above $1/7$ of its length. If the wave gets too steep, it breaks, and the energy goes into other waves. As a storm approaches a beach, the wave period lengthens, so instead of 10 small waves per minute, now 5 large waves strike the beach.

Waves move at 3.5 times their period in seconds, so a ten-second wave moves at 35 miles per hour. The longest recorded storm swell had a period of 22.5 seconds and a speed of 78 miles per hour. This is how severe storms send warnings to distant beaches 2 to 3 days in advance.

As a rule of thumb, the height of a wave will not be greater than half the speed of the wind. A 100 mile-per-hour wind will raise waves with a maximum height of 50 feet. The biggest wave on record was measured by the U.S.S. *Ramapo* on February 7, 1933. Seventy-mile-per-hour winds blew over an exceptionally long stretch of the Pacific Ocean. The waves were 15 seconds apart, and the biggest wave was 112 feet high.

The old belief that every seventh wave is higher has a limited amount of truth to it. Wave trains from distant parts of the oceans are converging upon the beaches. Depending upon the frequencies generated by the distant winds, waves will reinforce at regular intervals.

Winds and air temperatures play an interesting part in the life of freshwater lakes. If a south wind is blowing in the early spring, the north end of the lake will have warmer deep water, and the fishing could be much better in these warmer waters.

After the ice melts in the spring, lake water reaches its maximum density at 39°F, before it begins its spring turnover. As it warms, the lake begins to stratify into three layers if it is deeper than 60 feet. A lake with a depth of 60 feet will have three layers of about 20 feet each. The *epilimon* is the warm surface layer, followed by the *thermocline*, where the temperature drops about 1°F for every 2 feet. The *hypolimnion* layer at the bottom of the lake stays a constant 39°F year-round.

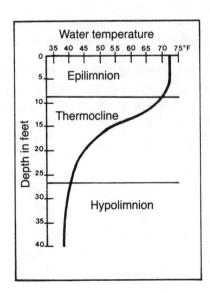

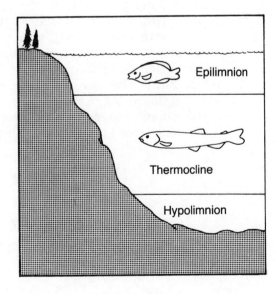

These layers are of great interest to fishermen because fish have different preferences for temperature, oxygen, and pH (measure of acidity or alkalinity). Fishermen sometimes use special reel thermometers to probe the depths of the lake. Below is information on the preferred feeding temperature of some common fish. Below a certain temperature the fish feed poorly, above a higher temperature, they cease feeding.

	Crappie	Lake Trout	Salmon	Sunfish	Trout
Lowest temperature	60°F	40°F	50°F	65°F	48°F
Best feeding temperature	68°F	41°F	61°F	72°F	61°F
Highest temperature	75°F	50°F	68°F	80°F	68°F

Winds add extra oxygen to the water, especially when whitecaps occur. Water will hold up to 2.6 percent oxygen at 86°F and 4.9 percent at the freezing point. Fish prefer areas with plenty of oxygen, and they often gather in places where a cold stream empties into the lake to find extra food and oxygen.

It appears that fish can sense the pH of the water. When fish in an aquarium were given a choice between water with an ideal pH and water with a low oxygen content, they took the ideal pH. The blood of fish, like ourselves, is slightly alkaline. Dissolved carbon dioxide in the water regulates the pH by combining with calcium in the water. Since pH varies throughout a lake, simple, treated paper strips can be used by fishermen to determine the best fishing areas. When a lake becomes too acidic, because of acid rain, fish can no longer adjust, and they begin to die.

Many ocean species prefer a certain water temperature. They follow warmer water northward in the summer and then follow it southward in the fall. Tuna boats keep track of the ocean temperatures, for tuna fishermen know they won't catch any fish in water that is more than a few degrees above or below preferred temperatures.

The rich cod fisheries off the coast of Iceland are also dependent on temperature. The fish stay in the *mixing layer*, which is normally 230 to 280 feet downward. Iceland normally has a stationary low-pressure center. If this should change to high pressure for any length of time, it disturbs the mixing layer, and the cod fishing is poor.

THE SAILOR'S WEATHER

"If it was my case, I shouldn't [go to sea], Captain Clubin. The hair of the dog's coat feels damp. For two nights past the sea-birds have been flying wildly round the lantern of the light-house: a bad sign. I have a stormglass, too which gives me a warning. The moon is at her second quarter; it is the maximum of humidity. I noticed today some pimpernels with their leaves shut, and a field of clover with its stalks all stiff. The worms come out of the ground today; the flies sting; the bees keep close to their hives; the sparrows chatter together. You can hear the sound of bells from far off. I heard tonight the Angelus of St. Lunaire. And then the sun set angry. There will be a good fog tomorrow, mark my words. I don't advise you to put to sea. I dread the fog a good deal more than a hurricane. It's a nasty neighbor, that."

—VICTOR HUGO, *TOILERS OF THE SEA*

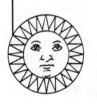

Ancient legends state that man did not sail the seas during the Golden Age. In early times, there was a belief that the wind belonged to God, and using it for boats was contrary to divine will. Survivals of this belief continued up until recent times. When combining equipment was introduced into Scotland, the clergy warned farmers against using "the Devil's wind" to separate the wheat from the chaff.

When William the Conqueror was ready to sail to England, there were no winds. After waiting a few days, his sailors began to gossip that God was signaling His disapproval. The winds came, and William conquered England.

Much of the old folklore of the sea concerns luck charms to get the wind to blow, or to prevent it from destroying the ship. In Homer's *Odyssey*, Aeolus bound the winds in a sack and released Zephyr for Ulysses. Seamen prayed to Castor and Pollux to deliver them from storms. St. Paul's Biblical shipwreck happened on a ship whose sign was Castor and Pollux.

Sailing the Mediterranean was a seasonal occupation, for few sailors dared brave the harsh winter storms, Ship owners charged higher rates in winter, but distances were short; navigation wasn't a problem.

About 270 years before the time of Christ, Aratus put together a book of weather signs, which was meant for sailors. All of the classical signs that mark storms are noted; from the ring around the moon, to changes in sound and apparent heights of the mountains. By noting these signs, the captains could head for a safe port during the winter sailing months.

When the first sailors ventured out beyond the pillars of Hercules (the Rock of Gibralter), they could no longer rely on seasonal winds or navigation

117

by nearby landmarks. If clouds and stormy weather hid the sun, what was one's direction?

The first Europeans to develop a partial answer to the weather problem were the Vikings. By A.D. 900, they traveled to Iceland. The first sighting of Iceland was made after Naddod was blown off course by a storm. The first ship to Iceland was piloted by Floki, on a voyage with no compass in an area that is often overcast.

The earliest tool that made ocean voyages possible on overcast days was the sunstone. One early sea saga reads: "The weather was thick and stormy. St. Olaf the king looked about and saw no blue sky. Then the king took up the sunstone and held it up and then he saw where the sun beamed from the stone."

The sunstone appears to be the mineral cordierite. The crystalline structure of this mineral polarizes light, and when it is rotated to a 90° angle to the sun, it changes color. The sunstone gave the Vikings the sun's direction in times of bad weather.

The idea for the compass came from China at the time of Marco Polo's journey to that ancient land. The first compass was a piece of magnetite tied to a piece of wood that was floating in a bowl of water. A sailor had to hold the bowl of water steady in the pitching ship to get the direction.

The true mariner's compass was first used around the year 1300 in the Mediterranean. It reached Scandinavia around 1350 and made the dangerous trip to Iceland much safer in bad weather.

The chief instrument of the sailors' weather forecasts was their knowledge of the winds. Shifts in its direction and changes in its speed foretold the weather. Every sailor knew this maxim: "A backing wind [counterclockwise] says storms are nigh, but a veering [clockwise] wind will clear the sky."

Sailors recognized that certain latitudes were storm generators, and that other areas were places where men might run out of fresh water waiting for the winds. For example, you might think the shortest distance between two points is a straight line. But if you tried sailing directly from Seattle to Hawaii, you might be stranded out on the ocean for months without winds. If you sailed a thousand miles down the California coast, and then straight to Hawaii, there is generally ample winds at all times.

John Josselyn was the first person to update the old weather lore and publish it in English. During his trips to New England, he made notes of the weather signs used by the sailors and farmers and published them in a book in 1663. The good weather signs are as follows:

> Moon fair and bright
> Distant hills are clear
> One rainbow follows a storm
> Mists in the valleys at night

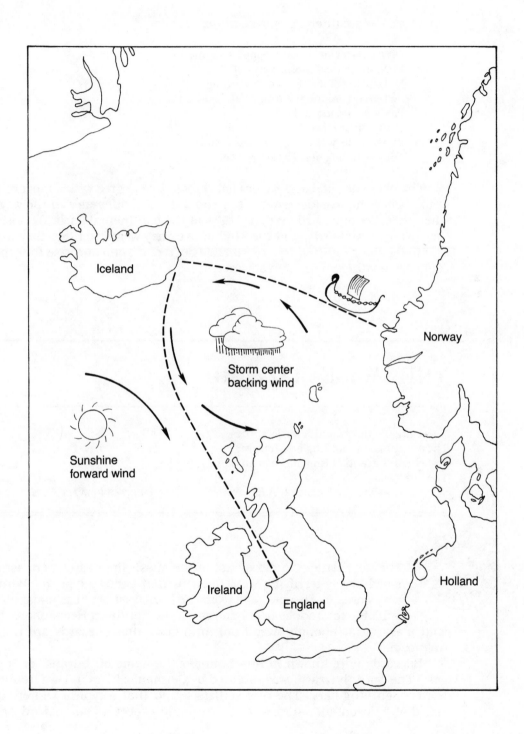

Iceland

Norway

Storm center
backing wind

Sunshine
forward wind

Ireland

England

Holland

Josselyn's bad weather signs were these:

> Obscuring of small stars: tempest approaching
> Changing winds: storm warning
> Murmuring of the sea: storm warning
> Red circles around the sun: wind
> Pale sun and moon: rain
> Distant hills black and cloudy: rain
> Distant hills yellow: certain sign of snow
> Sunrise hiding in dark clouds: rain

The old fishermen and sailors felt the sea to a degree which most of us would believe impossible today. They could feel the difference in the waves when their boats passed over the edges of the continental shelf. By noting the currents, seaweeds, and the kind of fish they were catching, they were able to tell their location to a reasonable degree. It is no surprise that they could feel weather changes in advance.

THE SAVAGE BLIZZARD

These are the terror of the settlers, these,
The arctic blasts, howling and freezing,
For which the land of England has no name—
The blizzards.
—WILLIAM HADDOCK, *A REMINISCENCE: THE PRARIES OF IOWA AND OTHER NOTES*

The first settlers to move west of the Mississippi River were astonished by the terrifying winter storms that swept the plains. Winter cold waves in Europe come from the Northeast, so little moist air is available to produce heavy snows. In the Southern Hemisphere, the land mass is far enough away from Antarctica that blizzards are nearly unknown.

Blizzards were known in Russia under the name of "burans" or "purgas." The word "blizzard" was invented by German settlers in Iowa from the words "lightning-like." The new settlers found that a typical winter consisted of a dozen snowstorms, but they could expect a real blizzard every fifth winter. A blizzard is defined as a storm with heavy snows, winds above 35 miles per hour and temperatures below 20°F.

During the winter of 1856–57 the settlers faced huge snow drifts with a crust of ice that would support men and wolves, but not grazing animals. Deer and the elk were trapped in the valleys where they starved to death or were killed by wolves and men. The elk never returned to Iowa, and it took nearly a century for large numbers of deer to return its wooded river valleys.

On January 23, 1886, the Kinsley, Kansas, newspaper printed a humorous article about two Santa Fe passenger trains that were stalled in blizzard snowdrifts. ". . . For three days that seem like an eternity (several eternities, in fact) the party have subsisted entirely on the crude polar berries and mosses dug out of the snow with incredible labor, and so fierce have the pangs of hunger become that in many instances the passengers have been compelled to devour the very crackers, cheese and spiritous liquors which they had brought with them for strickly medicinal purposes."

Blizzards are uncommon in the Eastern United States, but the storm of March 12, 1888 will never be forgotten. The groundhog didn't see his shadow that spring and baseball players were beginning training. The New York department stores chose that Monday to advertise as "Spring Opening Day." Walt Whitman, the staff poet of the New York *Herald*, caught the spirit of the mild spring weather by contributing a poem ending with, "The Spring's first dandelion shows its trustful face."

The Friday before the storm was the warmest day of the year, and temperatures soared into the fifties. That day a Herald reporter ran a story on the buyer for Ridley's department store, who had just purchased a cheap carload of snow shovels. The purchase was satirically called "Meisinger's Folly."

On March 8, a small storm started in the Salt Lake City area and began moving Eastward. At the same time, another storm was born in the Gulf of Mexico and it crept over the southern states. The Weather Bureau did not exist until 1891, but Army Signal Corps meteorologists were keeping track of both storms via telegraph. Neither storm seemed serious and Sergeant Elias Dunn telegraphed newspapers from his office in Washington, D.C. with the report that New Yorkers could expect "a cloudy Sunday followed by light rain and clearing."

Wind and rain came to the city on Sunday evening, and during the night, the rain turned to snow. By early Monday morning the air was thick with snow and the winds were gusting up to 84 miles per hour. The fierce winds snapped every power and telephone line in the city and there was no way of communicating with other cities. After trading a mere 15,000 shares of stock, a handful of Wall Street brokers tried to go home. The few banks that were able to open ordered all loans extended for there was no way they could do business.

The fierce winds blew 20 inches of snow into drifts up to 15 feet deep, and many frozen people were pulled from the huge drifts. The winds were so strong that people were blown back time after time when they tried to cross the streets, and they had to form lines in order to cross. The few horse-drawn cabs willing to make their way through the clogged streets charged high fares. Bartenders served "car driver's drink" which was a mixture of red pepper and ale. It was a time when people invited strangers into their homes, and on the streets people were rubbing ears and noses to thaw them out. Good humor was everywhere and signs were posted on snowdrifts "Keep off the grass" or "Don't pick the flowers."

Monday was normally the harbor's busiest day, and hundreds of ships waited to dock after a Sunday layover. About 200 ships were wrecked and hundreds of people died at sea that Monday. Seamen froze to death while standing watch in the icy winds and seas.

After 3 days of complete isolation, the first trains arrived on Wednesday with the aid of thousands of snow shovelers. On Thursday high temperatures filled the streets with melting slush, and the city went back to work again. There had been higher winds, more snow and colder weather in New York City, but never before had they all been combined. The utilities wisely buried the power and telephone lines in the following years, to prevent a repeat of the disaster.

Midwest winter storms are described as "cold waves," "northers," and "blizzards." The relative differences are a matter of wind velocity, direction, and the amount of snow. Cold air does not contain enough moisture for heavy snows, so warm moist air must be displaced in order to create a blizzard.

America's pioneer of meteorology, William Redfield, wrote in 1846, "It will be readily seen that on the approach of a great storm from the lower latitudes by the usual routes, while revolving from right to left, its first effect will be to bring in the warm and humid air of a more southern region, and when the axis of the gale has passed, the contrary result necessarily follows." This is what happened in the New York City blizzard. The weather reporter wrote, "The storm from the West has the arctic on its heels, and the disturbance from Georgia was sopping with warm moisture from the Gulf of Mexico. You couldn't find a better combination for concocting a howling blizzard!"

Western ranchers use snow fences to control the effects of blizzard snows. Huge drifts pile up directly behind the fences and protect roads that would otherwise drift shut. They provide shelter for the animals and water during the summer.

During World War II, Russian generals ordered the road to Finland held open at any cost. Large groups of Finnish laborers were conscripted to erect snowfences. The laborers began to erect the fences about a hundred feet away from the road, but the suspicious Russians ordered them placed next to the road. When the first snowstorm came, the road was blocked for the winter.

Blizzard conditions produced several interesting stories of the wild west. In October 1846, five feet of unseasonably early snow blocked George Donner's party north of Lake Tahoe in the Truckee mountain pass. While they waited, an eight day blizzard piled up drifts up to fourty feet. Trapped and low on food, the party was reduced to cannibalism. Five months later they were rescued, but only 47 of the 87 people survived.

In 1873, a party of gold seekers traveled into the San Juan Mountains of Colorado during the winter. Months later, only Alfred Packer returned, and when stips of flesh were found near the bodies of his companions, he was convicted of cannibalism and sentenced to be hung. At the trial, the judge was reported to have said: "Packer, you so-and-so, you have eaten half the Democrats in Hinsdale County." He was eventually pardoned, and today the lunchroom at the University of Colorado is named the "Alfred Packer Memorial Cafeteria."

During my youth, two feet of snow driven by 60 mile per hour winds whipped up eight foot drifts in Northeast Iowa. Blinding snow reduced visibility to near zero, and freezing rain put a crust of ice on the giant drifts. The snowplows were helpless under these conditions. Eventually rotary plows had to be sent from Colorado to clear the roads. The feeling of being helpless in a real blizzard is an experience that you don't forget.

V.
WEATHER AND WILDLIFE

THE WEATHER FISH

He knows the course of the stars and can always orient himself; he knows the value of signs, both regular, accidental and abnormal, in good and bad weather; he distinguishes the regions of the ocean by the fish, the color of the water, the nature of the bottom, the birds, the mountains and other indications.

—THE *MU'ALLIM* OR *PILOT OF THE ARABIAN SEA*.
THE BOOK WAS WRITTEN IN SANSKRIT
BY AN UNKNOWN AUTHOR IN A.D. 434.

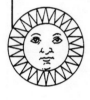 Fish live sheltered from the sights and sounds of the world of air, so it is not readily apparent why they should be weather sensitive. Most of this sensitivity may be due to atmospheric pressure changes, for this would affect a fish's sense of depth.

Although most fish go deeper during storms, some rise to the surface. New Zealand has a curious deep-sea fish known as the frost fish. It comes to the surface during periods of frost. During storms, New Zealand beaches are littered with fish.

The first people to link the weather with fish were the Egyptians. They believe the *boori* or mullet predict wind changes, because a shift of wind could leave the fish stranded in muddy water. If fishermen stop catching mullet on one side of a lake, they look for a shift in the wind.

Although we don't usually hear it, fish do a great deal of talking. It is known that the freshwater drum fish stops its talking on days when the temperature is low and the skies are cloudy. The repetition rate of the toadfish is water-temperature dependent. This fish sounds like the whistle of a boat. During World War II, many a toadfish committed suicide while seeking shelter alongside an acoustic mine.

The first United States Navy ship to visit Cambodian waters anchored off the mouth of a river in 1820. Lieutenant John White wrote that the men were astonished by hearing the sounds of an organ, bells, a guttural frog and a giant harp. The natives told them it was only the fish talking.

Fish talk to attract mates and socialize. The gurnard of the Mediterranean apparently talks about the weather. It grunts when storms are coming, and knowing fishermen head for port.

There are many accounts of fish being earthquake sensitive, particularly in Japan and China. Fish are often seen swimming erratically around the surfaces of ponds and rivers. Deep-sea fish that are rarely caught are suddenly caught in large numbers before earthquakes.

Japanese researchers tested the ability of the catfish *Parasiluros asotus* to predict coming earthquakes. The Asamusi Marine Biological Station kept

127

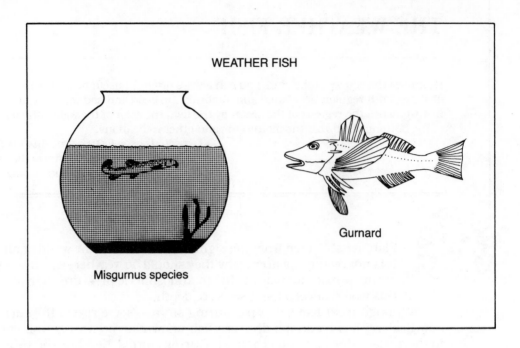

WEATHER FISH

Gurnard

Misgurnus species

catfish in tanks fed with water from the earth from October 1931 to May 1932. Several times a day, they tapped on the sides of the aquarium and noted the reaction of the fish. If the fish suddenly darted in alarm, they wrote "earthquake in 8 hours" on the blackboard. Their predictions were 80 percent accurate.

There are several accounts of fish swallowing rocks before storms. Fishermen often examine the contents of the stomach of their catch to see what they were feeding on. If they found rocks in the stomach of the codfish, the old-time fishermen would head for port. They believed that the fish were taking on ballast before the storm.

The best-known weather fish was the loach of European lakes. The word *loach* comes from the French word meaning "to fidgit." In good weather, this fish normally lies on the lake bottom. Before bad weather comes, it swims around on the surface or digs excitedly into the sand. Linnaeus and the older naturalists called the loach the "temperature fish." This isn't correct, because water temperature changes little.

Many people in Europe kept this small (2-to-3-inch) fish in an aquarium. It needs a fishbowl with several inches of fine sand and a water plant to give it some shelter. It should be kept in the shade in an even temperature and fed insects and worms. Its behavior changes several hours before storms come to the area, and it is generally reliable.

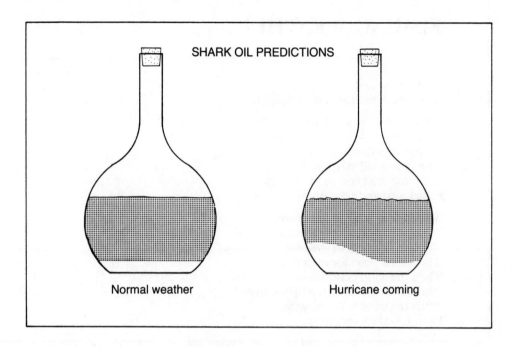

SHARK OIL PREDICTIONS

Normal weather   Hurricane coming

If you don't have a fishbowl, then the best thing you can keep is shark oil. There is an old belief in the Caribbean that "the shark knows where the weather is." The shark is very sensitive to vibrations, and could well sense coming hurricanes hundreds of miles away.

The best shark oil for hurricane detection is said to come from the liver of the puppy shark. Shark-oil barometers are sold in Bermuda for the purpose of hurricane forecasting. As the hurricane approaches, the oil is said to become cloudy from the direction the storm is coming. The top part of the oil forms little rivulets as the storm approaches, and this is a gauge of the intensity and nearness of the storm.

FISHING WEATHER

South wind and sky bright,
Gives the fisherman much delight.

Fishermen in anger froth,
When the wind is in the north;
For fish bite the best,
When the wind is in the west.
When the mist creeps up the hill,
Fisher, out and try your skill.

When the wind is in the north,
The skillful fisherman goeth not forth.
When the wind is in the east,
'Tis neither good for man or beast.
When the wind is in the south,
It blows the bait into the fishes' mouth.
When the wind is in the west,
Fishing's at its very best.

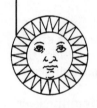

Going fishing is its own reward, but it is more rewarding to bring something home. Often, there is no explanation why the fish bite one day, and the next day nobody can catch anything. Fish generally feed early in the morning and late in the afternoon, but if food is available, fish can be caught at any time of the day, if they are biting.

Most fishing legends center on the direction of the wind. But those who have kept track of the catch and the wind claim that the wind is not a factor. During days of bright sunlight, fish react to the light and are difficult to catch. If wind is rippling the water, it diffuses the light, and fishing is likely to be better.

Sir Francis Chantry, the famous English sculptor of the nineteenth century, was fond of expounding his fishing weather theory. He always carried a thermometer, and if the water was warm and the air was cold, he went fishing. He felt that at such times fish were active and insects were not; therefore, the fish would be hungry.

The American Indians said, "Fish go down deep, means a coming storm." Fish do move into deeper water as the barometer drops. Their swim bladder is filled with air, giving them a neutral buoyancy. As the pressure falls, the fish would tend to rise. In order to counterbalance the decreased pressure, they go several feet deeper.

In 1937, the Sportsmen's Club of Lincoln, Illinois, made a study of fishing and barometric conditions. The fishermen found that when the barometer was above 29.90" and rising or steady, the fishing was good. When the barometer was at a low point or falling, the fishing was poor. Falling barometers mean stormy conditions, and rising barometers generally mean good weather. These conditions are ideal for fishermen, too.

The Lincoln study found that the wind had no effect on fishing. Whenever there was a prolonged low-pressure system, the fish that were caught had empty stomachs. Trout and bass usually feed near the surface, and they were the most sensitive to a low barometer. Walleyes feed on the bottom, and they were less sensitive to barometric conditions. During a low-pressure period, it was best to fish with worms, minnows, and live bait in deep waters. When the pressure was high, it was best to fish near the surface with lures.

A study was done in Kansas by surveying fishermen and then comparing their answers to barometric readings. This study seemed to indicate that the best fishing took place during a steady medium-high barometric reading. On a very low barometric reading, 1 fisherman would be reporting good results, while 5 would be reporting poor fishing. For a steady middle barometric reading, the ratio was 1 to 3, while at very high readings it dropped to 1 to 6.

There is a widespread belief that fish don't bite when it thunders, and one proverb goes; "When trout refuse bait or fly, there is a storm nigh." These statements may be true on occasion, but at other times, the best fishing is during pre-storm conditions. Lakes appear to get worse before storms, but rivers and streams have a brief period of good fishing as the water begins to rise. Perhaps these beliefs reflect the frustrations of the fishermen.

The largest freshwater fishing prize in North America is the muskie. This fish is difficult to catch, although it may bite any time of the day. A computer was used to analyze the weather conditions on 550 muskie catches. They were caught more frequently in south and west winds, and infrequently in north and east winds. They were rarely caught when there was no wind or when the skies were overcast. They moved into deeper water during bad weather, and few were caught.

The greatest influence on the time to go fishing came from the "solunar theory" of John Alden Knight. He got the idea from the Seminole Indians, who called it "Up sun, down moon." When the moon is directly overhead, it is high tide. Six hours and 12 minutes later, the tide is at its lowest point. The small fish that feed in shallow waters are forced into deeper water, and the big fish have a feast.

Knight introduced the theory of solunar fishing in 1935, and fishermen have been using it ever since. A number of experts claim that positive thinking is involved; you expect to catch fish and you do. Many people believe that it does have a measurable effect on the catch, and Knight received thousands of letters from satisfied fishermen.

The solunar theory is not a weather theory, but it is related to the cosmic influences that drive the weather. A good fisherman will catch fish if they are there to be caught at any time, but some of us need all the help from our knowledge of the weather that we can get.

There is an old story that some successful fishermen have watched the goldfish bowl in order to know when to fish. When the aquarium is churning with hungry fish, grab your pole and head for the lake!

That story might be the result of another cosmic influence. On a dark night in 1945, a number of boats were out on a Wisconsin lake, but nothing was happening. A brilliant aurora lit up the night sky, and suddenly the lake came to life with jumping fish. The fishermen began to catch fish, and many of them got the limits in a short time. Then the aurora faded, and the fish quit biting. The northern lights may be both a weather influence and a clue to good fishing.

HUNTING WEATHER

All the creatures had to stay inside, because day after day the South Wind blew strongly. They couldn't hunt, their eyes smarted, and they were hungry. Raven called a war council with orders to defeat the South Wind.

Despite a strong wind, the animals reached the home of the South Wind. The mouse bit the nose of the South Wind, and he dashed out and slipped on the fish outside the door. The other animals beat him with clubs, and in order to save his life, the South Wind agreed to give them 4 days of good weather mixed with 4 days of bad weather.

—THIS IS ONE OF THE STORIES THE INDIANS OF THE NORTHWEST
TOLD TO EXPLAIN WHY THE NORTHWESTERN COAST HAS SO MUCH RAIN.

Through the ages, the weather instincts of both fish and animals may have developed through the survival of the fittest. The animals that didn't react or learn from weather conditions perished, and the others survived. Many bird species and caribou might be said to have developed the greatest weather reactions, for they migrate.

Since deer react to strange scents, they are usually gone before hunters get close. Deer move into the wind in order to receive warnings about any

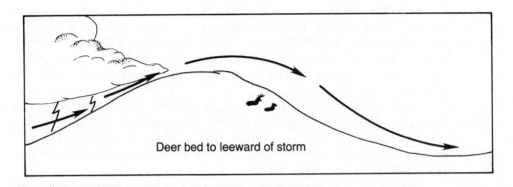

Deer bed to leeward of storm

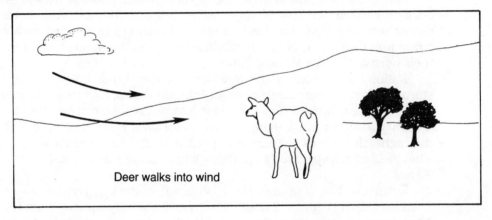

Deer walks into wind

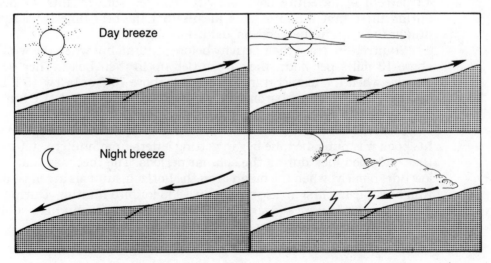

Day breeze

Night breeze

danger near them. In order to utilize this, the hunter must move against the wind to surprise his quarry. The hunter can determine the wind by watching clouds, falling leaves, and swaying bushes.

The deer is also a creature of sound, and hunters who walk steadily are warning the deer. The Indians used to say, "Walk a little, look a lot." A deer does just that; it takes a few steps, looks around, and then continues. They notice too much noise or too little noise. The best hunting time is often when there is a gentle rain or a little snow. The leaves don't crackle, and the snow subdues noise.

Deer are used to snow, drizzle, and cold, but they dislike storms. During a high wind, they can't hear warning sounds and they can't locate disturbing scents. During storms, they choose sheltered areas such as dense river brush, pine forests, and the lee side of mountain ridges. It is very difficult to find them then.

Just before a storm arrives, they go out to feed, because they might not be able to feed for several days. After a storm passes, deer come out everywhere and feed. The best times for hunting are just before a severe storm and during the clearing conditions of a winter snowstorm. Deer lose their normal caution at these times.

Rabbits also react to the weather, for they must hole up when it is stormy. When the barometer rises and the sky clears, rabbits are out feeding again. The best time to trap rabbits is just before a storm, because they are not wary at that time. During the day, rabbits are rarely seen, but during the night, thousands of them come out to feed. I have seen a Texas winter wheatfield at night under a spotlight where there was a rabbit every 10 to 20 feet.

Squirrels, too, have had their behavior studied in relationship to the weather. During the first 2 hours of daylight, the gray squirrel runs through 43 percent of its entire day's activity. The fox squirrel hits 47 percent during these first 2 hours. This means that the best time for a squirrel hunter to be out in the woods is just before daylight.

Squirrels increase their activity before a storm, but when the wind gets above 13 miles per hour, they are curled up in their holes. They become inactive when the temperatures rise much above 70°F or below 15°F in the cold months. When the barometer is above 30.2″ and the relative humidity is 60 to 90%, it is an ideal time for them to move.

John Alden Knight formulated his solunar theory for fishing, but when his book was published, he began getting letters from hunters stating that they had good luck during the solunar periods. The theory is that during low tide, namely, when the moon is on the horizon, animals are most active. Knight wrote his first article on this in 1936 for bird hunting. During most of the day, birds were so wary that they flew before the hunters could get close, but during the solunar periods they were less cautious.

Knight received a letter from a deer hunter stating that he had shot a buck on a particular day. He asked Alden, "What time of day did I shoot this animal?" Knight looked at his tables and sent a reply to the hunter. The hunter sent a card by return mail; "You are right!" That of course, is an ideal case, but there are times when we have "good luck."

How did the Indians hunt big game with bows that had rarely more than 30 pounds of pull? They understood the habits of animals and they shot from very close ranges. They knew that the winds shift during the day. During the day, winds flow uphill as the sun heats air, but in the cool of the evening, they drift downhill. Given a choice of paths, animals walk into the wind. Thus, the hunter goes into the wind for maximum success.

Deer and other animals choose protected areas during stormy conditions. Often these are just below the crest of a hill, where the wind spills over and turns backward. With an understanding of feeding habits, it is possible to track animals and surprise them.

One group that really relies on the weather is duck hunters. Duck weather consists of spitting rain, fast-moving low clouds, and gusty winds. The ducks are thinking, "Let's go south fast." They can't fly high on account of the low clouds, because they use rivers for navigation. During the nights, they look for grain fields that they can feed in before starting the long, tiring journey.

For those of us who are "camera hunters," our ability to get pictures depends upon the weather as well as our feeling toward life. Normally reclusive birds can be seen before storms, because our feathered friends lose their caution before bad weather.

INSECTS AND WEATHER

If ants move their eggs and climb, rain is coming anytime.
When eager bites the thirsty flea, clouds and rain you sure shall see.
When spiders' webs in air do fly, the spell will soon be very dry.
Low over the grass the swallows wing, and crickets, too, how sharp they sing.

—WEATHER PROVERBS INVOLVING INSECTS

The first person to provide scientific information about the weather sensitivity of insects was the great French entomologist Jean Fabre. He wrote a set of books on his study of insects. As an example of his observations, he theorized that spiders didn't stick to their webs because they had an antistick compound on their legs. When he washed a spider's legs in a solvent, it stuck fast to its own web.

The pine precessionary caterpillar browses on pine needles during the cold months of December and January. Each night, Fabre's caterpillars left their nest around 9:00 P.M. and returned at 2:00 A.M. One night, he invited a forest ranger over to watch the caterpillars go out and feed. That night, the caterpillars stayed at home, puzzling the two men.

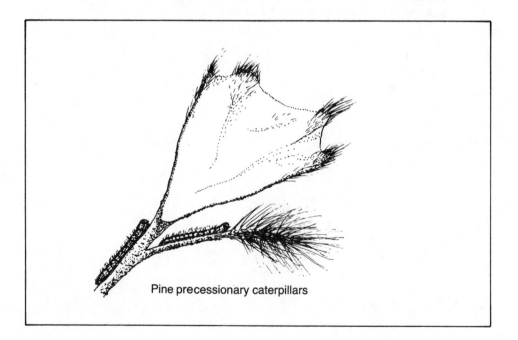
Pine precessionary caterpillars

A snowstorm blew in that night, and from then on Fabre began to check weather conditions against caterpillar behavior. When the barometer dropped or there were cold north winds, the caterpillars stayed at home. When the barometer rose, they went out and fed. Fabre consulted his caterpillars when he planned a trip into town to pick up supplies.

Fabre also kept colonies of geotrope beetles. These are the "dung beetles," which were used as the sacred scarab symbol in ancient Egypt. It provides for its larvae by burying animal droppings as food for them. In order for cattle ranching to be successful in Australia, these beetles had to be imported, because cattle manure remained unburied and stunted the growth of grass. These beetles become very restless and excitedly buzz their wings whenever the barometer drops, thus providing Fabre with another storm warning.

Many bits of folklore involve spiders. When spiders take down their webs, obviously a storm is coming. When they build them, good weather is on the way. Pliny the Elder mentions that spiders know that a river is about to rise, for they begin to build higher webs.

The legend of the spider as weather prophet is due mainly to a book by Denis Quatremere-Disjonval in 1797 called *On Spiders.* Some spider beliefs are as follows:

> Complete orb webs: dry, sunny weather
> Eating old web: rain coming
> Building radii in bad weather: clearing 12 hours away
> Complete orb web: rain will stop quickly

On October 14, 1922, the New York *Herald* ran a story called "A Trusted Missouri Weather Prophet." It seems that Will Brown had an old spider that he used as his weather forecaster. When the Macon, Missouri, fair organizers were unable to get a week-long forecast from the weather bureau, they went to Will Brown. He told them, "When a spider runs out slender filaments, it is a sure sign of fair weather for at least a week." And his prediction proved to be right.

East Coast American sailors believed that spider webs in the rigging were a sure sign of a southwest wind. Sailors along the Amazon River believed that spider webs in the mast meant a storm. I believe that both beliefs were true, but that the spider webs were actually there all the time. As the relative humidity rose, the webs trapped moisture, and the sailors could see them, so they did indicate rain.

Dr. Henry McCook, the American spider expert, kept spiders for 6 years and observed their weather senses. None of his spiders had any predictive value. Both Jean Fabre and other spider experts have come to the same conclusion; spiders can't predict the weather. They operate on seasonal timing, and they do pretty much the same things every day at the same time. If they happen to be taking down their webs and it rains, they are just lucky. Much of the time, they will spend hours in pouring rain on their webs. It may be that there are some weather-sensitive spiders among the 30,000 species, but all of the spiders that have been studied are not.

Many beekeepers believe that bees have an acute weather sense and signal the presence of bad weather by changing their buzzing rate. There are two theories as to how bees sense changes in weather. One theory holds that ice crystals preceding storm systems scatter light, thereby confuse the directional mechanisms of bees' compound eyes. The other theory involves electricity: Bees carry a static electrical charge of about 11 picocoulombs (pC) at a humidity of about 70%. As the relative humidity rises to 90%, the charge drops to less than 1 pC. When bees depart from their hives in the morning, they are negatively charged, but when they return, they carry a positive charge of up to 1.8 volts. Pollen is negatively charged and sticks to the bee's body by electrostatic attraction. Bees become restless and irritable when exposed to negative ions or when their hives are near high-voltage lines. Their irritability may be due to a lack of ability to carry pollen to feed to the young bees as the humidity rises. A poet described their weather sensitivity this way:

> When bees to distance wing their flight,
> Days are warm and skies are bright.
> But when the flight ends near their home,
> Stormy weather is sure to come.

The Australian aborigines believe that the activity of the mound-building termites is an indication that wet weather is coming. By adding a high addition to their mounds, they will be able to move into it during heavy rains. In a recent case, the Australian government stopped logging operations when the prediction of Guboo Ted Thomas came true: that heavy rains would bring severe erosion in logged-over areas. He based his prediction on the activity of the termites.

In November of 1885, an English journal printed a letter saying that the termites were very active in Brisbane, Australia, and mentioned the aboriginal belief that this predicts rain. I went through the old rainfall records to see if the termite prediction had come true. During the next two months Brisbane had 17 inches of rain, the heaviest in 6 years. The termites were right!

In China, the hum of the cicadas is the basis of a proverb: "Hearing the cicadas in the rain foretells the coming of fine weather." When the relative humidity gets too high, cicadas can't vibrate their wings. Often the humidity drops in the tail of a storm, even while a few drops of rain are still falling, and the cicadas began to hum.

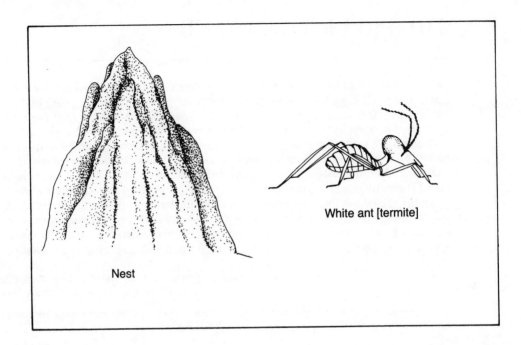

White ant [termite]

Nest

Before hurricanes arrive, flies become nervous and seek shelter, then fall into a sort of coma. Mosquitoes attack more readily and then go into hiding. Cockroaches start rushing around in broad daylight.

It appears that all of this behavior is related to the drop in barometric pressure. When the pressure returns to normal, the insects act normally. There may be some influences from negative ions on insect behavior as well.

One of the most interesting findings of scientists is that caterpillars hatch into moths or butterflies only during drops in barometric pressure. If the pressure drops suddenly, the cocoon will hatch a few days early or up to a week late. At first appearance, it would seem rather traumatic for the new butterfly to experience a storm. But that is nature's provision for using high winds to fling the new insects far away from the territory claimed by its parents. In this way, southerly winds are able to take butterflies far to the north and enable them to exploit new sources of food.

THE PREDICTIVE LEECH

He told us how that he had got two favorite leeches. He had been blooded by them last autumn when he had been taken dangerously ill at Portsmouth; they had saved his life, and he had brought them with him to town, and ever since kept them in a glass, had himself everyday given them fresh water, and had formed a friendship with them. He said he was sure they both knew him, and were grateful to him. He had given them different names, Home and Cline (the names of two celebrated surgeons) their dispositions being quite different. After a good deal of conversation about them, he brought them out of his library, and placed them in their glass upon the table. It is impossible, however without the vivacity, the tones, the details and the gestures of Lord Erskine, to give an adequate idea of this singular scene. He would produce his leeches at consultation under the name of "bottle conjurors" and argue the result of the cause according to the manner in which they swam or crawled.

—A LETTER BY SIR SAMUEL ROMILLY ABOUT THE ENGLISH CHANCELLOR LORD ERSKINE AND HIS ECCENTRIC HABIT.

One of the major theories of medicine that slowly died away in the nineteenth century was that sickness was due to bad blood. In the year 50 B.C. a Greek doctor from Laodicea introduced the leech to medicine, and it became an instant success. In order to get rid of bad blood, you had leeches suck it out of you.

The leech is a segmented worm related to the earthworm. It possesses a sucker at each end and lives by sucking blood. It attaches itself and punctures the skin with its three teeth and injects an anticoagulant (hirudin), which allows the blood to flow freely. When it has sucked 10 to 15 milliliters of blood, it drops off.

The medical leech is an interesting harmony of browns, greens, blacks, and yellows. During the eighteenth century, ladies used to embroider leeches into their dresses or wear leechlike markings. Because of their widespread medical use, it is not suprising that people watched their activity in relationship to the weather.

The great English songwriter William Cowper, who wrote such favorites as "Amazing Grace" and other well-known church hymns, remarked that "leeches are worth all the barometers in the world." Other people were beginning to note the weather activity of leeches.

One of the first accounts of leech forecasting published in a science journal was written by Dr. Henry Season in 1806. He kept his leeches in a glass jar covered with a piece of cloth. He changed the water once a week in the summer and every two weeks in the winter. I have put his observations into modern English.

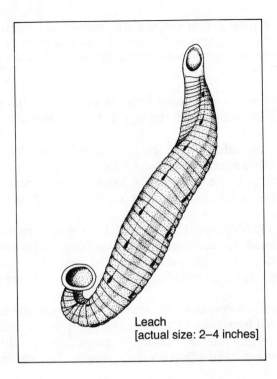

Leach
[actual size: 2–4 inches]

If the weather proves to be serene and beautiful, the leech lies motionless at the bottom of the jar and rolls up into a ball.

If it rains, either before or after noon, it creeps to the top of the jar and remains there until the weather changes.

If wind is coming the prisoner gallops through its habitation with amazing swiftness and seldom rests until it begins to blow.

If thunderstorms are coming, it remains continually outside the water for a day or more making throwing and convulsive motions.

In frost as in clear summer weather, it lies constantly at the bottom.

In snow as in rainy weather it lives at the top of the jar. Whatever the reason for this, I must leave philosophers to determine.

In 1816, James Stockton had an article printed in the *Annals of Philosophy* on leech predictions. He says that he observed leeches for a long time in relationship to the weather and comments:

In fair and frosty weather it lies motionless and rolled up in a spiral form at the bottom of the glass; but prior to rain or snow it creeps up to the top, where if the rain will be heavy, or of some continuance, it remains for a considerable time, if trifling it quickly descends. Should the rain or snow be likely to be accompanied with wind, it darts about with amazing celerity, and seldom ceases until it begins to blow hard. If a storm of thunder and lightning be approaching it is exceedingly agitated, and expresses its feelings in violent convulsive starts at the top or bottom of the glass.

Stockton remarks that before the heavy rains of July 1816, his leech was in constant motion. It descended from the top of the jar when the rains began.

A "Tempest Predictor" was exhibited at the great London exhibition in 1851. Dr. George Merryweather arranged 12 leeches in 12 jars so they could see each other and wouldn't get lonely. When the leeches rose to announce a severe storm, they tripped a mechanism that rang a bell, so everyone would be warned.

The leech was eventually forgotten in Europe, but it was rediscovered after the Chinese revolution of 1948. There was a new surge of interest in using ancient wisdom. Many communes established weather stations, where clouds, winds, and agricultural phenomenon were utilized for weather forecasting.

The chief tools were leeches, loaches, and snails. One commune found that the March forecast called for strong winds without exceptional weather. That day, the leeches escaped by pushing aside the gauze cover on the jar, and the weather fish were swimming wildly on the surface. That afternoon, a severe hailstorm hit the area.

On June 7, 1959, another commune had scheduled its field to be sprayed with insecticides. Rain would dilute the solutions and ruin the work. The cadres noted that the leeches were rising, the loaches were swimming on the surface, and the snails were sinking. The spraying was postponed, and heavy rain began to fall.

The ability of the leech to predict the weather has been forgotten ever since we stopped using them in medicine. They remain in the mud of swamps and lakes, keeping the secrets of weather prediction their ancestors have passed on for millions of years.

BIRD PREDICTIONS

. . . I rose on my feet, looked towards the southwest, where I observed a yellowish oval spot, the appearance of which was quite new to me. Little time was left me for consideration, as the next moment a smart breeze began to agitate the taller trees. It increased to an unexpected height, and already the smaller branches and twigs were seen falling in a slanting direction towards the ground. Two minutes had scarcely elapsed, when the whole forest before me was in fearful motion.
. . . Never can I forget the scene which at that moment presented itself. The tops of the trees were seen moving in the strangest manner in the central current of the tempest, which carried along with it a mingled mass of twigs and foliage that completely obscured the view. . . .

—JOHN JAMES AUDUBON, THE GREAT BIRD PAINTER, HAS PLENTY
TO SAY ABOUT THE EFFECTS OF THE WEATHER ON ARTISTS,
BUT NOTHING ABOUT THE EFFECTS ON BIRDS.

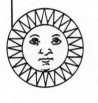

At the beginning of World War II, the Army Air Corps undertook a study of soaring birds in hopes of gaining information for troop-carrying gliders. A nest of vultures was captured and the chicks raised in captivity, for vultures are the heaviest and most efficient of soaring birds.

When the birds were old enough to fly, the scientists set up their instruments and released them. Day after day, the young vultures walked around as the frustrated scientists waited. In desperation, one of the men began making short runs, flapping his arms vigorously, while the vultures looked on. After watching the scientists, the vultures caught on and began flying as they were supposed to. You might ask, if men have to teach birds how to fly, what can birds possibly know about the weather?

Since Noah released his raven and dove from the ark, men have believed that birds have special abilities to predict rain. There are several birds that utter calls before it rains. One French saying has it that "when the wood-pecker cries, it announces rain." This bird calls *plui, plui,* which in French means "rain, rain." The robin is noted for flying about and uttering loud chirps before rain. The rain quail of India gives its musical whistle call during the hours before rain.

Other birds remain silent before storms. On August 15 and 16, 1898, there was an exceptionally severe storm in Illinois. When people in the area were surveyed on their observations on the storm, many noted that the birds stopped singing two days in advance. William Warner remarked, "For forty-eight hours before the great electrical storm, not a sound was heard

Green Woodpecker

Rain Swallow

Wilson's Storm Petrel

from our wild birds." He had been in Ceylon 5 years earlier and had asked his guide, "Where are the birds today?" The guide replied, "Big storm come." A severe lightning storm came. When it was over, the birds resumed singing. In March 1886, Warner traveled to the Kilauea volcano in Hawaii. The birds in the area were silent, and soon there was a series of earthquakes and an eruption. Then the birds began to sing again.

One of the strangest observations of birds comes from Norway. One day in 1949, the adult swifts disappeared, and the young swifts went into temporary hibernation. The next day, there was a severe electrical storm. After it passed, the adults returned. Nobody knows where the adults went.

The behavior of hawks has long been used to predict hard rainstorms. If you travel in the country, you might see them perch on telephone poles to get a good vantage point before heavy rains come. During dry weather, mice and voles nest in low ground and creek bottoms. These creatures began moving toward higher ground before storms, and as they cross the roads, the hawks catch them.

The Japanese have a series of weather proverbs based on bird behavior. Hawks fly in pairs through the mountains before rainstorms. Rabbits and mice are also moving from their hiding places at this time. The blue magpie normally hides itself in trees. When it comes out and can be seen, it is likely to rain.

There is a Japanese proverb that warns: "Wear your straw raincoat if kites are flying in the morning, but take off your hat if you see a kite in the evening." The birds are also believed to predict typhoons. The critical time for typhoons in Japan is during the first two weeks of September. People feel relieved if jays are around; the absence of jays is taken as a warning sign.

Fishermen used to gauge the behavior of sea gulls to see if it was safe to go out to sea. If the sea gulls were hanging around on the beach, it seemed a good idea for the fisherman to follow suit.

Seed-eating birds can stick around during the winter, but an insect-eating bird has to migrate in order to survive. Robins and bluebirds are known as *weather migrants*, because they travel northward as soon as it seems warm. Other birds are called *calendar migrants* because they arrive in spring on a certain date.

Migrating birds time their spring flights to coincide with the ending of a high-pressure system and the southerly flow of wind in a low-pressure storm center. Bird watchers are able to predict migratory behavior by watching the high- and low-pressure systems.

During the fall, birds begin their migration in times of high pressure when visibility is good. Day travelers, such as swallows, require sun fixes to take their position. Night travelers can orient themselves by the stars. Most birds have to have good weather in order to determine their position, and they tend to postpone travel until good weather returns.

Bad weather has been responsible for spreading birds to many countries. In 1937, flocks of fieldfares were heading from Scandinavia to England, when freak winds carried them to Greenland. The survivors were able to start colonies there and live. Our feathered friends may have other secrets hidden in their bird brains, and perhaps our weather senses will be developed by studying them.

ANIMAL FORECASTS

My Lord be praised with all your creatures,
Most of all brother sun;
It is your day and light he brings us.
For he is lovely, shining with great splendor
And of you he brings us tidings.

My Lord be praised by sister moon and stars,
Lovely, clear and precious, you shaped them in the sky.

My Lord be praised by brother wind,
By air in every weather, bad or good,
It is from them all your creatures gain their food.
—Saint Francis, "Hymn to All Creation"

It is generally believed that animals know when storms are approaching. Many proverbs speak of this. "When the ass begins to bray, we will have rain that day"; "When horses rub their backs on the ground, rain is sure to come around"; "When goats begin to snort, then comes weather of another sort."

The general restlessness of animals prior to a storm has been observed by many farmers. Cattle eat extra food before storms, for instinct tells them they may not have food for days. Pigs begin building nesting areas. They carry straw and pieces of wood to form shelters. This behavior was written about by Pliny the Elder almost 2,000 years ago.

Other proverbs deal with the fact that animals are more likely to stick close to their shelters if severe weather is coming. Welsh weather lore states: "If cattle and sheep go up the hill, the farmer is glad; if they come down, he is anxious."

Along Interstate 5 near Roseburg, Oregon, there is a peak called Mount Nebo. There is an old belief that the wild goats grazing on the mountain predict the weather by their actions. If the goats go high, good weather is forecast for the day; if they stay low, bad weather is portended. The Roseburg radio station has given goat forecasts. When these were checked against weather bureau predictions, the goats lead by 90% to 65%.

A Huntsville, Texas, paper held a contest between the weather bureau and forecasting based on traditional observation. John McAdams used his cow Brimmer to predict the coming weather. Brimmer held an early lead, but on the final day she failed to predict rain. Final score: weather bureau, 19 points; Brimmer, 15 points.

GORDON WIMSATT'S BEAR GREASE INTERPRETATIONS

Clear

Clouds coming

Clouds at hand

Clouds from
direction of point

Clouds from
two directions

Double front
coming in

Future storm

Light wind

Strong wind

Mixed winds,
future wind

Hurricane
starting

Tornado

Several
tornados

Earthquake
coming

Earthquake

Volcano
eruption

Atomic
explosion

Traditional lore attributes cats with the power of prediction. If the cat looks out the window or scratches the table legs, it will rain. A pharmaceutical institute in Zurich, Switzerland, evaluated a group of 9 cats on their weather behavior. They found that there was a great deal of difference in behavior between individual cats at the same time. The cats slept more during warm, high-pressure weather and became restless as the weather changed. They were most active during times when it was cool and humid. If you are testing tranquilizers on cats, and they are tranquil, is the drug working well? Not necessarily. It may be the weather.

There is a passage in the log of a British ship in the Persian Gulf that indicates how cats occasionally do sense coming storms in a dramatic way: "The first indication of the approach of the storm was the behavior of the ship's cats. For the previous weeks they had been very lazy and sleepy but about 12 hours before the storm, they went quite mad, rushing wildly about and biting people's feet."

The naturalist Hal Borland noted that much of the old animal lore didn't seem to have any validity. He did note that muskrats would suddenly start digging new holes higher up on stream banks. On these occasions, heavy rains followed.

The Indians looked at the bear as a medicine man and a weather prophet. Since it's difficult to have bears around a camp, the Apache Indians filled animal bladders with bear grease. The clear tissue enabled them to watch the cloudlike patterns forming in the grease, and these patterns were used to predict weather.

Gordon Wimsatt of Cloudcraft, New Mexico, has carried on this method of weather forecasting for over 50 years. He is called upon by ski operators to see if snow is in the offing and by farmers looking for rain. His patterns are even said to pick up atomic blasts. He has been featured in newspaper articles and on television shows.

Mr. Wimsatt holds the theory that animal cells respond to the weather even after the animal has died. He has had remarkable success in predicting unusual weather. The day after "tornado patterns" showed in his jar of bear grease, 8 tornadoes hit the Midwest. He uses a wide variety of animal fats, but bear grease works best.

Texas ranchers have used rattlesnake weather lore. "If you see a lot of rattlesnakes, it will rain," they maintain. That might be true, for cooler weather precedes rain, and rattlesnakes come out in cool weather. "If rattlesnakes are heading for high ground, buy cattle" is another proverb. That could be true. As for the bit of lore that states "Rattlesnakes are sluggish during droughts and vicious in bad weather"—I wouldn't want to be the one to find out.

VI.
BOTANICAL WEATHER

PLANTS AND WEATHER

After the French revolution, a "calendar of reason" was adopted on October 5, 1793. The months were named after weather or natural occurrences, beginning with the Autumn equinox on September 23.
  Autumn months: *Vendemaire, Brumaire, Frimaire*
  Winter months: *Nivose, Pluviose, Ventose*
  Spring months: *Germinal, Floreal, Prairial*
  Summer months: *Messidor, Thermidor, Frucidor*

English wit inspired Sheridan's rhyming calendar, which also began in the fall.
  Autumn months: Wheezy, Sneezy, Freezy
  Winter months: Slippy, Drippy, Nippy
  Spring months: Showery, Flowery, Bowery
  Summer months: Hoppy, Croppy, Poppy

Plants must grow and reproduce in all kinds of weather conditions. There are a number of proverbs indicating that plants can be used to predict the weather. What do they really know?
     The scarlet pimpernel is known as "the poor man's weather glass" in England. It closes its petals when the relative humidity reaches 80 percent. By closing its petals, it prevents wet weather from damaging its pollen. Tests show that it lives up to its poetic reputation:

> Pimpernel, pimpernel, tell me true,
> Whether the weather be fine or no;
> No heart can think, no tongue can tell,
> The virtues of the pimpernel.

The dandelion is also known as a weather prophet, but it does not react to changes in humidity. At temperatures below 51°F, it folds its flowers. It could be predictive of cold weather in the spring or autumn.
     It is a popular belief that clover contracts its leaves at the approach of a storm. Clover folds its leaves when wind velocity is higher than 20 miles per hour. High winds are often found before rain, but clover protected from wind does not fold its leaves before a storm.
     The ox-eye daisy (*Chrysanthemum vulgare*) also folds its flowers at relative humidities that range from 64 percent to 82 percent. During tests, the flower was once observed folding up at a relative humidity of 54 percent, and an hour later a shower dampened the field.
     Because flower petals fold with a rise in humidity, they can be used to make hydrometers. The petals of *Acroclinium roseum* have a very sensitive

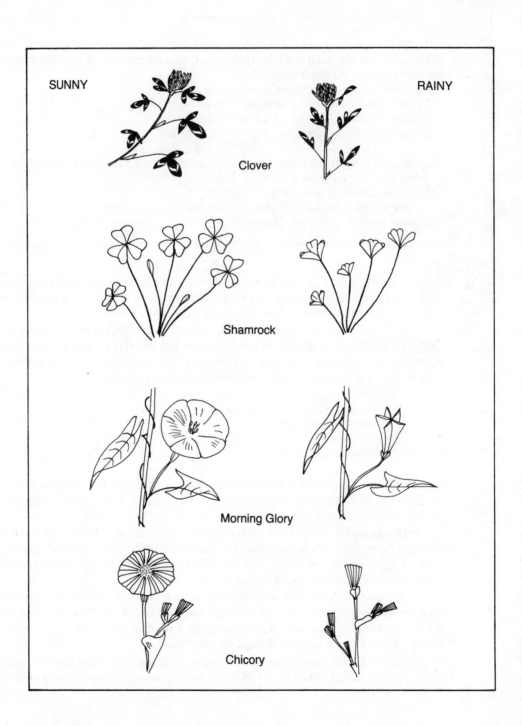

SUNNY   RAINY

Clover

Shamrock

Morning Glory

Chicory

hinge mechanism. A hair is glued to a petal to act as a pointer, and then the petal is glued to a card. Calibration points can be made by comparing it to a standard hydrometer.

The Peruvian plant *Porlieria lorentzii* folds its leaves at night and expands them in the morning. It folds its leaves more slowly on a sunny day than a cloudy day. If it folds up its leaves half an hour before sunset, the next day will be fair. If it folds them an hour before sunset, it will be overcast and stormy.

Hipolito Ruiz made some notes about this plant when he did a study of Peruvian vegetation two centuries ago. He kept two plants as weather observers for three months. He was satisfied by their performance, but he did not leave behind a complete account of his experiment.

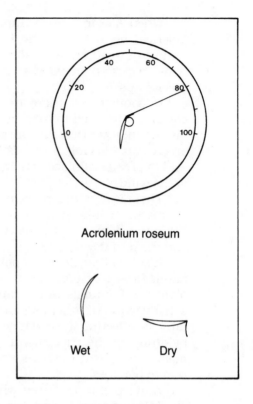

Acrolenium roseum

Wet   Dry

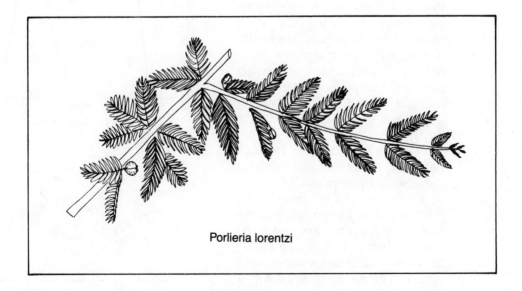

Porlieria lorentzi

When Alexander von Humbolt made his geographic surveys of South America, he looked at the *Porlieria*. "The closing of its leaves indicates change of weather, generally the approach of rain," he wrote. "This plant is more certain in its tokens than any of the Mimosacea, and it rarely deceived us."

In popular lore, there are several "rain trees." The cottonwood and the poplar are believed to turn up their leaves when rain is coming. My observations indicate that they turn up their leaves whenever it is dry and slight breezes are blowing, and this has little to do with rain. The silver maple is said to predict rain accurately by turning up its leaves, but this may be due to the wind.

There is a "wind and rain" flower, which grows in China. The flowering of this plant is said to be a sure indication of rain. When an American delegation of meteorologists visited China in 1970, the flower was blooming, and the next day it rained.

It seems that the lowly potato outstrips all of these weather plants. Dr. Frank Brown devoted most of his life to studying the secret rhythms of plants and animals. A rat can be studied by putting it in a cage with an activity wheel. If it is more energetic on certain days, then an outside force must be affecting it. Brown began by studying the rhythms of animals and then turned his attention to potatoes.

A live potato is a seed which slowly burns oxygen, and this rate of "breathing" is the key to what the potato is "feeling." The potatoes "feel" the moon, for they burn more oxygen during the dark phase of the moon and less at the new moon. Dr. Brown put potatoes in sealed containers with constant pressure, temperature, humidity, and light. There was no way these potatoes could "feel" anything.

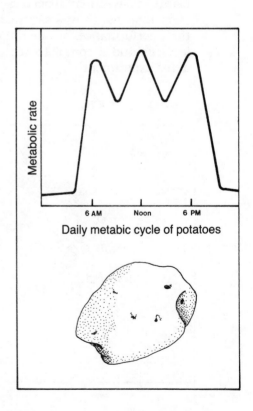

Daily metabic cycle of potatoes

During each day, Dr. Brown's potatoes generated an activity curve with three peaks: one each at 6:00 A.M., 12:00 noon, and 6:00 P.M. Dr. Brown worked for several years and accumulated 1.25 million potato hours before he knew what the potatoes were saying. The morning peak was a measure of how fast the barometer had changed on the previous day. The noon peak was related to the outside temperature. The real puzzle was the evening peak. Eventually, it was found that the peak anticipated barometric pressure two days in advance. Since barometric pressure is directly correlated with weather conditions, the potato "knows" the weather!

Dr. Brown believes potatoes are sensitive to high-energy background radiation. They cannot react to cosmic rays because they are blocked by the blanket of air. Cosmic rays do affect the electrical conductivity of the atmosphere. There is a 27-day periodicity in potato respiration, indicating that the potato is sensitive to the sun.

It is unlikely that potatoes will appear on the 6:00 P.M. news with their impressions of future weather. In any event, the measurement of a potato's metabolism requires special equipment, presently available only to highly trained scientists. It appears that the predictive potato will have to serve as French fries for the present time.

THE WEATHER PLANT

Greetings in cold, windy climates:
  English: "Good morning."
  French: "*Bonjour.*"
  German: "*Guten tag*"
Greetings for hot, dry climates:
  Egyptian: "How do you perspire?"
  Arabic: "May God strengthen your morning."
  Persian: "May God cool your age."
Greetings for a hot, wet climate:
  Venezuelan tribes: "How have the mosquitoes used you?"

Charles Darwin was the first scientist to study the movement of plants. After his return from the voyage of the *Beagle*, he lay on his couch, sick from an unknown disease, and noted the movements of his plants. Nearly all plants move during the day, but their movements are so slow, we don't notice them.

We do notice one plant's movements, for it has the distinct appearance of a nervous person. *Abrus precatorius* is a member of the bean family that grows wild in India. On any normal day, it can be seen raising and lowering its branches and folding and unfolding its leaves. If you touch a leaf, the entire branch will begin folding up. The plant folds up at night or when rain or hail touches its leaves.

The plant became known as the "weather plant" in 1887, when Josef Nowack applied for an English patent. He couldn't patent a plant, but his application stated, "Certain conditions must be observed in order to cultivate Nowack's weather plant in such a manner than it can be used as a weather indicator. . . . We have constructed an apparatus which in combination with the weather plant constitutes the principal subject of our inventions."

The newspapers began to print accounts of this wonderful plant. An English paper wrote:

> The weather plant continues to excite considerable interest at Vienna. Men of science, who on its first discovery were unwilling to venture an opinion on its prognosticating virtues, now agree, after extensive experiments that the shrub is in truth prophetic. Thirty-two thousand trials made during the last three years tend to prove its infallibility. . . ."

The *London Times* printed a letter in which Mr. Nowack's agent stated:

> The observatory of the Austrian Tourist's Club on the Sonnwendstein at an altitude of 1,511 meters in the Styrian Alps, well known to many English tourists, which supplies the various branches of the club with weather forecasts during the season, has now for already over a year, discarded both aneroid and ordinary barometeres for that purpose and depends for its forecast upon the weather plant alone.

The weather plant was just too good to be true, and Mr. Nowack was invited to the Royal Botanical Gardens at Kew, England, to deliver his bonanza to mankind. He was introduced to the gardeners by a letter from the Prince of Wales, and experiments upon the fabled plant began.

The plant was housed in an inverted glass bell jar with temperature remaining relatively constant. Nowack divided the plants into B (barometer) plants and T (temperature) plants. The T plants went to sleep the quickest in the hot afternoons. The plants were aligned to the points of the compass. If the young leaves from the north side exhibited unusual movements, then thunderstorms were supposed to occur in that direction two days from the present time. If the older leaves exhibited unusual movements, then a local thunderstorm would happen in two days. If the rachis moved, the branches vigorously up and down, then an earthquake was indicated.

Nowack soon added the direction and strength of the wind, the amount of atmospheric electricity, snow, and fog to his predictions. He began drawing up barometric charts with hourly changes. Some charts were

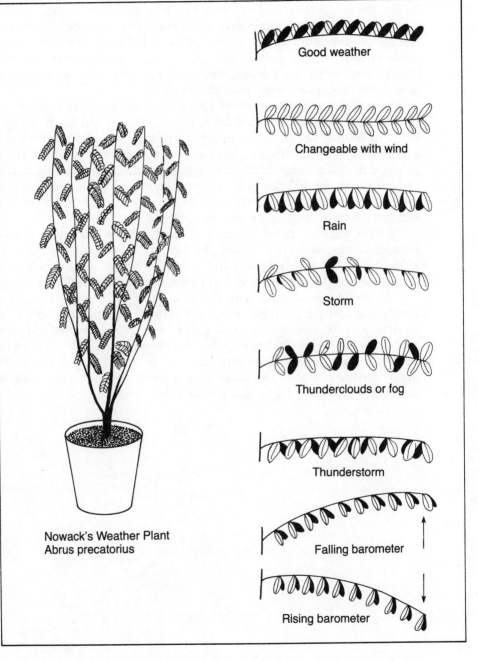

Good weather

Changeable with wind

Rain

Storm

Thunderclouds or fog

Thunderstorm

Falling barometer

Rising barometer

Nowack's Weather Plant
Abrus precatorius

accurate, for English weather is a mixture of rain and sunshine. His coworkers soon noticed that he selected his weather charts. Sometimes his plants predicted one day in advance, sometimes four days. Whenever Mr. Nowack found a weather chart that matched his weather-plant predictions, he put them together, and told everyone how accurate he was.

The earthquake predictions made from the up-and-down positions of the rachis were equally vague. Any shift in light and darkness resulted in some shifting of the plant's branches. Often this occurred when clouds passed over. Although Nowack claimed to have predicted earthquakes years ahead, there is no evidence that this is so.

In spite of being denounced by English botanists as a fraud, work on the weather plants continued for another 20 years. In 1905, Nowack published a 94-page book in Germany. By 1908, he announced that he had found the most sensitive weather plants from all over the world. He planned to issue detailed 3-day advance weather forecasts and 28-day forecasts for critical times for fire damp explosions in mines. No articles appear after 1908, and no books were written until his death in 1918, so I presume the weather plant is keeping its secrets to itself.

If the weather plant actually worked, the person who should have noticed it was the great Indian plant physiologist Jagadis Bose. He, too, worked with the weather plant and noticed changes in its oscillations when clouds passed. He did not find that these oscillations predicted the weather.

There may be another way of investigating the weather plant and the weather sensitivity of other plants. Philip Callighan attached a sensitive electrometer to the spines of a plant. The plant reacted differently on mornings before thunderstorms hit the Florida area. After thunderstorms, his plants had electrical spasms during the night, when it was normally quiet. Perhaps plants do know more about the weather, and with an electrical read-out, we'll be able to learn their secrets.

THE OAK AND THE ASH

Ash when green, is fire for a queen.

"Avoid an ash, it courts the flash."

Every oak has been an acorn.

Oaks may fall when reeds brave the storm.

—TREE PROVERBS

One of the best-known English weather proverbs reads: "Oak choke, ash splash." The proverb refers to the early time in spring when the leaves are coming out. If the ash is first, rain will come (splash); if the oak is first, then dryness (choke) is in store for the summer.

The study of phenology encompasses the blooming of cherry trees and the appearance of the first birds, insects, and butterflies. Records have been kept in England for two centuries about common natural events and the dates on which they occur each year. A study of these events reveals natural weather patterns and variations in spring temperatures. This could be of value to farmers, because they must plant crops that will mature in time for fall harvesting.

The oak and the ash have been the subject of more weather proverbs than any other plants:

Oak before ash, have a splash.
Ash before oak, have a soak.

Oak before ash, there'll be a splash;
Ash before oak, there'll be a choke.

When buds the oak before the ash,
You'll only have a summer splash:
When buds the ash before the oak,
You'll surely have a summer soak.

If the oak is out before the ash,
'Twill be a summer of wet and splash.
But if the ash is before the oak,
'Twill be a summer of fire and smoke.

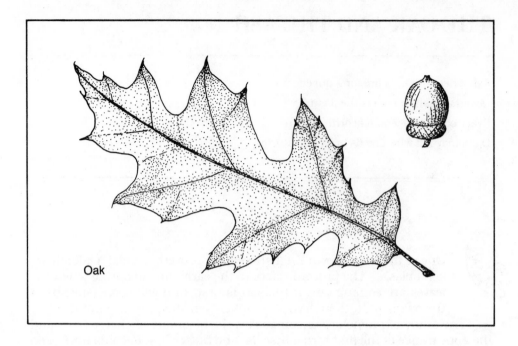

Oak

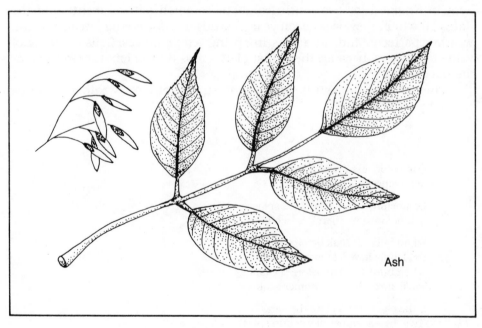

Ash

The following proverbs put it differently:

> If the oak before the ash comes out,
> There has been or will be a drought.

> If the oak is in leaf before the ash,
> 'Twill be dry and warm, and good wheat to thrash.
> If the oak and the ash open their leaves together,
> Expect a summer of changeable weather.

From all these proverbs, you can choose the simplest variations, which are "Oak choke, ash splash" or "Oak smoke, ash squash." You can interpret them oppositely, so you can choose whichever variation fits the year. This proverb reflects the problem of nature observations. A shaded oak leafs later than an ash exposed to direct sunlight. Oaks begin to leaf from the top branches; lower branches leaf last.

There are long lists covering the dates of the leafing of both oaks and ashes and other phenological observations. When I compared a 60-year sequence with the leafing dates, I noted the wet and dry years. When I took the four wettest summers and looked at the leafing calendar, there was no deviation from normal for either oaks or ashes.

The *London Times* published a letter on May 6, 1954, in which the writer noted that the ash was far behind the oak in leafing that year. It had been a dry spring, and he quoted "Oak before ash, in for a splash." It was dry for the next two months, then England had the wettest summer and fall in 50 years. If he had quoted "Oak choke, ash splash," he would have been wrong. The endurance of these proverbs is a sure justification that you'll always be 50 percent right when you quote them.

PRACTICAL PHENOLOGY

When mulberry trees are green,
Then no more frosts are seen.

When the elm leaf is like a mouse's ear,
Then to sow barley never fear.
When elm leaves are as big as a shilling,
Plant kidney beans if you are willing.
When elm leaves are as big as a penny,
You must plant kidneys if you want any.

Cuckoo oats and woodcock hay,
Make a farmer run away.

—THE MEANING IS THAT THE FARMER CAN'T SOW OATS UNTIL THE
COCKOO COMES, WHICH IS GENERALLY AROUND APRIL 14.
THE WOODCOCK VISITS ENGLAND LATE IN THE FALL WHEN IT HAS
GENERALLY BEGAN TO RAIN; THEN IT IS TOO LATE TO MAKE HAY.

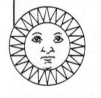

In 1863, Charles Morran coined the word *phenology*, which he defined as "the science of appearances." Phenology is a record of the time of the melting of the ice, the first dandelion, the first cricket, and the first frost. It is a record of the appearances of flowers, the leafing of trees, the return of the birds and the insects.

The first practical use of phenology is found in the eighth century B.C. The Greek writer Hesiod wrote, "Men may sail in the spring when the top leaves of the fig are the size of a crow's footprints." At that time, the dangerous winter storms were over in the Mediterranean, and the sea was safe for traders and fishermen.

We derive our word *month* from the old word for *moon.* In ancient times, people used the lunar calendar, for there was no paper and often no clear understanding of the 365-day year. Since the lunar calendar had to be constantly corrected, farmers used phenology for planting and harvesting. Julius Caesar reformed the calendar in 46 B.C. and fixed the months in relationship to the appearance of stars on the horizon, thus producing a true solar calendar.

Although calendars are fixed in time, seasons are not. We divide our year into 4 seasons, but the farmer knows that spring comes when it is time to plant crops. The difference in spring between southern Alabama and the Canadian border is about 2½ months. Spring moves 1° northward in 4 days and arrives 1 day later for every 100 feet of altitude.

The Chinese were using phenology for their farmers as early as

2000 B.C. They planted and harvested crops in relationship to the signs given by wildflowers and trees, for the solar calendar was unknown.

Grape vines should be pruned during the dormant season so that the wound will heal by the time the sap begins to run in the spring. Late pruning damages the vitality of the vines. During Roman times, farmers who pruned their grape vines too late were called "cuckoos." Although the remark was derisive, it was the time when the cuckoo was returning to Italy.

With the changing time of spring, how did the North American Indians know when to plant corn? John Heckwelder wrote, "When the leaf of the white oak is the size of a mouse's ear, it is time to plant corn. When the whippoorwill is hovering about calling 'wekolis,' it is to remind them that it is time to plant corn."

The settlers adopted a name from the natives as a useful part of phenological lore. The blooming of the shad bush (*Amelanchier canadensis*) was the signal to go out and catch shad. Shad came up the streams and rivers to spawn then.

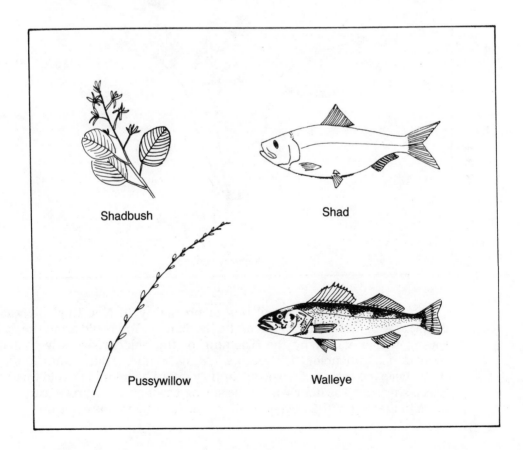

Shadbush

Shad

Pussywillow

Walleye

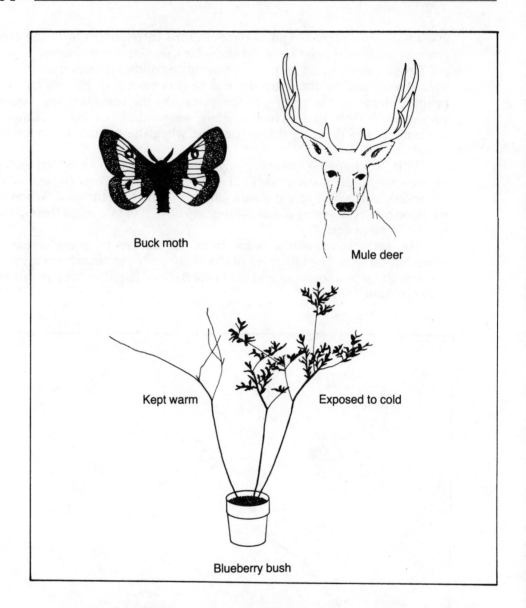

Buck moth

Mule deer

Kept warm   Exposed to cold

Blueberry bush

One of the most practical bits of phenology in New England was a simple way to avoid the ravages of the Hessian fly. You could avoid the fly by sowing winter wheat by the blooming of the goldenrods. New England farmers used phenology for sheep shearing. Farmers didn't want to shear their sheep too early in the spring, or they might be injured by cold weather. Sheep shearing was done when "the spring-sown grain begins to carpet the fields in green" or "the wool goes off, as the fruit blossoms come on."

The dates of the blooming of the cherry trees have been recorded for over a thousand years in Japan. Scientists have compared these dates to layers of snow taken from the Arctic icecap. During warm conditions, snow has more heavy oxygen (O_{18}) which has 2 more neutrons than O_{16}, which is normal oxygen. The records of the early blooming of cherry trees correlate exactly with the amounts of oxygen isotopes in the snow.

The fishermen of Japan have a tradition that it is bad luck to capture a sea lion when fishing for bonitoes which are a warm water fish. The truth of the tradition is that whenever sea lions appear, bonitoes are scarce, because sea lions like cold water. On the other hand, whenever sea lions are around, the sardine fishing is good because sardines are cold water fish.

The meaning of winter for many plants was discovered when growers tried to hybridize blueberries. They began growing new plants in heated, well-lighted greenhouses, but the blueberries refused to grow during the winter. Furthermore, the plants kept in a warm greenhouse refused to begin growing in the spring until months after plants exposed to the winter. A blueberry was grown with one branch outside the greenhouse and the other one in. The outside branch bloomed normally in the spring; the inside branch remained dormant.

The scientists theorized that the plants needed winter in some way. They believe that the freezing action cracks some cells and releases an enzyme that turns starch into sugar and begins the growing process. It is for this reason that walnuts will not grow unless they are put into a freezer and kept for several weeks, before being planted.

Fishermen can use phenology to indicate good times for going after certain fish. The budding of pussy willows in the north coincides with the walleye spawning season, and it's a good time to catch them. When the wild onion blooms, crappies and bass begin their spawning seasons, and fishing is generally good at this time, too.

Hunters can use the signs of nature as well. The buckmoth is so named because it hatches when the deer are in rut. When the quaking aspens turn yellow, the bull elk begin to bugle and collect their harems. The wild turkeys become amorous when the redbud trees begin to bloom.

Nature has served as man's calendar for thousands of years. Our reliance upon the solar calendar has prevented us from observing many interesting relationships, which are ultimately based on the weather.

RAIN TREES, RAIN FORESTS

A Kansas farmer knew that it would be a very dry summer. He had just stocked his pond with fish and he didn't want to lose them when the pond dried up. In order to save them, he put them on the shore 10 minutes the first day, 20 minutes the next day, and so forth. By the time the pond dried up, the fish were eating with the chickens. Later that summer, somebody asked the farmer how his fish were doing. "Terrible," the farmer replied. "We had a severe thunderstorm and they all drowned."

—A TALL TALE.

Does the forest make the rain, or does the rain make the forest? The question sounds childish, for everyone knows that rain makes the forest. The areas of the world with heavy rains are covered with dense rainforest.

James Merriam was perhaps the first person in modern times to realize that forests can produce a rainy climate. He visited the island of Curaçao 60 miles from the coast of Venezuela. The remaining inhabitants told him that the island had been a lush forest, but after the trees were cut, it had become an arid desert.

When the trees surrounding Lake Valentia in Venezuela were cut, the water level rapidly dropped. Soon afterward, there was a civil war, and the inhabitants were killed or driven away. The trees grew again, and the level of the lake rose. When the trees were again cut, the water level began falling.

Jomo Kenyatta, the first president of independent Kenya, wrote about the tribal life of his people in *Facing Mount Kenya.* In his childhood, there were groves of sacred trees scattered across the countryside. The Kikuyu tribesmen used these groves to hold sacred ceremonies to Ngai, the Supreme Elder, and to pray for rain. Kenyatta recalled that each time the tribe held a rain ceremony, torrential rains fell.

The rapid spread of Christianity brought about a great change in the culture of the people. The groves of trees were cut to "destroy the influence of Satan." British planters were eager to cultivate the land "wasted" by the tribesmen. Kenyatta wrote:

> It is an undisputed fact that rainfall is much less in Kikuyu than formerly. I myself remember pools in which a full grown man could get out of his depth, and in which we all used to bathe, but which are now replaced by dry soil. The chief physical reason is in all probability the destruction of the forests.

The Panama Canal route goes through Gatun Lake, and the water for the locks comes from the streams feeding into it. In 1983, for the first time

in the canal's history, there was hardly enough water to operate the locks. Could this be due to the heavy cutting of the surrounding forests?

Plato wrote:

> . . . When Attica [Greece] was intact her mountains were heavily forested and the country produced boundless pasture for cattle. The annual supply of rainfall was not lost as at present by being allowed to flow over the denuded surface into the sea, and so was received by the country in all its abundance, into her bosom where she stored it in impervious potter's earth, and so was able to discharge the drainage of the heights into the hollows in the form of springs and rivers with an abundant volume and a wide territorial distribution.

Since Plato lived 400 years before the time of Christ, the forests of Greece must have been cut hundreds of years before. But history holds even more astonishing facts, for the barren, desertlike plains of Iraq were heavily forested 5,000 years ago. Remnants of that forest existed in 300 B.C. for Quintus Curtius, the Roman historian who wrote Alexander the Great's biography, stated that his route along the Karun River was heavily wooded the entire way. It is dry and barren today.

Moses was once given the Biblical promise of the "land of milk and honey." Israel is dry and arid, but to the north were the once-large forests of the "cedars of Lebanon." Did the cutting of these forests change the rainfall patterns?

The walls of Egyptian tombs are covered with hunting scenes of the pharaohs spearing lions and big game. But the Nile valley has no big game, and the Sinai is a dry desert. In that time, it was almost impossible to travel south of the second cataract of the Nile. Evidence has been found in Egyptian tombs that much of the timber needs were supplied locally. The names of a number of locations on the eastern desert and the Sinai bear the names of ancient trees.

Some of the Egyptian hunts must have occurred in present-day Libya. It is a barren country today, but in Roman times, it was forested. This brings up the interesting question of where Hannibal got his elephants. It is unlikely that he could have brought them by ship from southern Africa, and the Sahara was still far too vast to march these thirsty beasts across. The only answer is that elephants still lived in North African forests.

The disappearance of that forest is spoken of in an old manuscript telling the story of Sidi Tayeb, a Moslem holy man. He was bitten by a snake in the Guir (North African forest), and as he died, he called to the animals of the forest: "I order you at once to leave the region of the Guir which I take under my protection." Six days later, the animals were said to have left. At the bottom of the document was added, "And his followers to whom he bequeathed the forest cut it down."

The Romans were quite aware of the Guir, for in A.D. 47, Roman General Suetonius Paulinus wrote about the great forest that covered an area from

Libya to Morocco. No forests exist today; the land now serves only Bedouins grazing herds of goats. The Romans seem to have been aware of the need for forests, for Emperor Hadrian issued an order protecting trees and had it carved into stone. The stone survives, but the trees have long been cut down.

One of the most inspiring stories of our time was told by the French writer Jean Giono. He was asked by a group of magazine editors if he could create an unforgettable character. In response, he created the story of a simple shepherd in a barren area of France near the Italian border. His name was Elzeard Bouffier. Day after day, the shepherd planted 100 acorns. He didn't care that the land didn't belong to him; he was simply doing it for the good of all. As Elzeard Bouffier planted his acorns, rains returned, and the dry creeks filled with water. Flowers and vegetation returned to the dry area. As productivity increased, people moved into the area, and families were established. The story has such a strong grain of truth that it has been translated into several languages and has inspired a whole generation of foresters.

Rainfall and cloud generation depend upon rising and descending currents of air. Forests cool air and provide descending currents. It may be possible to optimize this process by planting carefully spaced forests. Forests were first planted in the Multan area of the Punjab in India around 1920. By 1940, for the first time within memory, regular rainfalls began.

The creation of the Sahara desert may be due to both the widespread cutting of trees for firewood and the appetite of goats. Goats strip the bark off trees and kill them. They eat vegetation so close to the ground that it dies out. Goats were an important resource for the few peasant farmers that had them in the beginning. In the end, they may be the prime factor in the extension of the Sahara desert.

Even in dry areas, like the Dust Bowl of the United States, it is possible to establish forests. The *Acacia peuce* is a large Australian tree that grows in true desert conditions. It is not the only desert tree which could establish forests in arid land. One of the driest areas of the world is the Desert of Atacama along the Chilean coast. If *Prosopis tamarugo* is started here, it will grow into a 60-foot tree watered only by coastal fogs. The seeds must be soaked in sulphuric acid for several minutes to make them sprout; then they are watered to get them started. It is an excellent construction wood, and plenty of potential forest land is available in the dry desert.

We study history because we believe that we can benefit from history. It is time to be concerned about the deforestation of the Amazon, the Sahel, and the slopes of the Himalayas. The Sahara is expanding, and the roots of the current famine might be found in the ceaseless cutting of trees. In the last 25 years, the percentage of forest area in Ethiopia went from 17 percent to 3 percent. Trees may be the difference between milk and honey, or sand and desert.

VII.
THE
WEATHER,
YOU, AND ME

HUMAN WEATHER REACTIONS

... When he comes to take a pleasure in making such observations, it is amazing how great a progress he makes in them, and to how great a certainty he arrives at last, by mere dint of comparing signs and events, and correcting one remark by another. Every thing, in time, becomes to him a sort of weather-gage. The sun, the moon, the stars, the clouds, the winds, the mists, the trees, the flowers, the herbs, and almost every animal with which he is acquainted, all these become, to such a person, instruments of real knowledge.

—FROM THE INTRODUCTION TO THE SHEPHERD OF BANBURY'S RULES
FOR PREDICTING THE WEATHER IN 1744.

There are days when we feel great, and there are days when we can hardly get out of bed. Is this due to the weather? The weather could have affected Abraham Lincoln's life. On a January day when the temperature fell by 50°F and a storm passed, Lincoln was so emotionally upset that he broke his engagement. Two weeks later, the same weather pattern occurred, and this time Lincoln was left rambling, incoherent, and threatening suicide.

When a study was made of the weather factors surrounding 527 suicides in Pennsylvania, it was found that neither sunshine or gray days affected the victims. The relevant factor was barometric pressure. Suicides rose by 30% whenever pressure changed by more than 0.35 inches in a day. A falling barometer was slightly more dangerous than a rising barometer.

The dark winter nights of the northern climates make some people overly depressed. It is believed that the lack of sunshine increases the production of melatonin; this additional hormone depresses people. The remedy is to get a full-spectrum sunlamp and spend several hours reading or relaxing under it daily.

Many zoo animals are from tropical regions, and they have similar winter problems. They become inactive and their circulation slows. During winter months, our relatives, the primates, are likely to lose weight, be listless, uncoordinated, and suffer from diarrhea. During these same months, humans have flu epidemics.

Hot weather changes our behavior, for we are either lethargic or hot-blooded. Most riots and revolutions occur in hot weather, because nobody feels like overthrowing governments in cold weather.

Shakespeare talked about hot weather: "I pray thee good Merat; let's retire; the day is hot, the Capulets abroad. And if we meet, we shall not escape a brawl, for now, these hot days is the mad blood stirring."

Anatole France wrote of hot summer weather, "What a noble people is

173

ours, it never makes revolts in winter. All the great revolutionary days are in July, August and September. When it rains they go home, taking their flag with them. They will die for the idea, but they won't catch cold for it."

A number of books and medical reports have dealt with the weather in relationship to medicine. Yes, we have more headaches, more heart attacks, and more psychic problems when the weather changes. In most cases, the results are about 5% above normal. I submit that what we are really looking at is stress. We are late, we have a harder time getting to work, and become more worried and fatigued. In spite of many medical reports, I don't believe that we're looking at significant weather factors, even though many experts think so.

The factor that does make a major difference in our weather feelings is the ionization of the air. In normal weather, there are approximately equal numbers of negative and positive ions. When strong winds blow across desert country, they produce high concentrations of positive ions. The Italians call this wind the *sirocco*; the Spanish call it the *solano,* and the French call it the *mistral*. The *Santa Ana* winds make the southern Californians crazier than normal. All of the reports suggest that winds with high concentrations of positive ions bring irritation, gloominess, and the blues.

The opposite condition is produced by negative ionization. We feel great when we stand at the base of waterfalls, and we sing in the shower. Summer showers produce a feeling of freshness and well-being.

Doctors have studied the effects of ions in people with sinus conditions. When volunteers were unknowingly breathing positive ions, they developed nasal obstruction, itchy noses, and husky voices in 2 to 3 minutes and their breathing capacity fell by 30%. When they breathed negative ions in the air, these conditions were reversed. These experiments suggest that anyone with breathing difficulties ought to purchase a negative-ion generator.

The most interesting results were obtained when researchers tested the effects of negative ions on young and old rats. Rat IQ is based on a measure of the time it takes the rat to run a maze correctly.

NORMAL AIR	NEGATIVELY IONIZED AIR
Young rats 26.5 errors in 22.8 minutes	27.5 errors in 23.8 minutes
Old rats 78.9 errors in 48.9 minutes	38.6 errors in 22.7 minutes

The experiments suggest that one of the best things we can do when we get old and forgetful is to stand under waterfalls or get a negative-ion generator. We will feel better, and our memory will greatly improve.

An interesting experiment was done to find out how people react to the weather. Students stood at convenient street locations and stopped people. They stated "I'm from the sociology department. We're conducting a survey of social opinions. Although the survey is eighty questions long, you don't have to answer them all." The answers were correlated with weather conditions. In the summer, people were more helpful when the day was sunny, with mild winds, and low humidity. People were less helpful on days with cold winds and during a full moon.

At the same time, a popular restaurant was surveyed for the size of patrons' tips. Sunshine and low humidity were two factors that induced larger tips. During good weather, women and older people tended to dine out, and they left larger tips, which may have influenced the result.

A final survey was done of people leaving the restaurant. "What did you think of the waitress?" they were asked. On sunny, mild days, people replied that the waitress was pleasant. During times of high humidity and a full moon, more people felt that the waitress was bad tempered.

WEATHER DREAMS

. . . People just chance to have visions resembling objective facts, for their luck is merely that of people playing games. This principle is expressed by gamblers. "If you make many throws, your luck will change," and that's always true.

It is not strange that many dreams are unfulfilled, for that's the same way with weather signs. If a stronger cause occurs, then the predicted event does not take place . . .

—ARISTOTLE, *PROPHESYING BY DREAMS*

. . . The same dream came often in my life, sometimes in one form and sometimes in another, but always saying the same thing. "Socrates," it said, "Make music and work at it." And I formerly thought it was urging and encouraging me to do what I was doing already and that just as people encourage runners by cheering, so the dream was encouraging me to do what I was doing, that is, to make music, because philosophy is the greatest kind of music and I was working at this. . . .

PLATO, *PHAEDO*

In ancient times, it was commonly believed that dreams were the key to inner worlds. During sleep, the soul could look ahead on the time track and study its future or solve the problems of everyday life.

The Polynesians ate a "dream fish" as part of a religious ritual in order to go into a sacred trance. The eater of this fish believed that he had died, experienced the afterlife, and returned with wisdom.

Dreams were carefully kept and interpreted in many of the ancient civilizations. The Biblical patriarch Joseph became a dream interpreter while he was kept in an Egyptian prison. It was during this time that the Pharaoh had a dream in which 7 fat cattle were devoured by 7 lean cattle.

In Egyptian symbolism, the Nile was often written alongside a symbol for 7 cows because farmers often grazed their cattle in the lush river grass. Joseph interpreted his dream to mean that 7 good years would be followed by 7 years of drought. The dream came true, and he received honor and rewards for having history's first recorded weather dream.

The Icelandic sagas contain many references to dreams, and these people were very concerned about weather-dream symbolism, for theirs was a harsh climate. In the Saga of the Volsungs, Gudron tells her maid that she had an unlucky dream. The maid replies, "Tell it to me, and let it not worry you, because one always dreams in bad weather." Gudron replies: "But this is not bad weather."

In chapter 34 of the saga, Kostbera says to her husband Hogni, "I dreamed a polar bear came in and broke up the king's high seat and shook his paws at us, so that we all grew frightened, and he had us in his paw at one time, so that we could do nothing and great terror prevailed." Hogni replied, "There will be a great storm, if you thought of a white bear."

The saga of Guomund also contains the story of how Bishop St. John appeared to people in dreams and assured them the weather would improve if his holy relics were exhumed. After they were dug up, the weather improved. In a climate such as Iceland's, you paid attention to weather dreams!

Lama Anagarika Govinda took a long Buddhist pilgrimage in Tibet and noted the old proverb "when you dream of dead people, it will rain." He found that his dreams of the deceased were inevitably followed by rain or snow in three days. Could we be subtly influenced by changes in barometric pressure, or do we become more sensitive to subtle vibrations before bad weather arrives?

It was once believed that the full moon made people "loony," and that during that period, your dreams would come true. The full moon does produce the "Transylvania effect" in many people. During that time of the month, your dreams show less REM (rapid eye movement), and the poorer quality of sleep could send a disturbed person over the edge. When admission times at mental institutions were surveyed, a full moon was not a factor. But hot, humid nights were a definite factor when people were admitted. The reason may be that a lack of sound sleep produces mental disturbances.

A survey was taken of a thousand medical cadets at the School of Aviation Medicine. They were examined for signs of drowsiness during a routine examination. This data was then plotted against barometric readings. When the barometer stood at 30.00 inches, more of the men were wide awake. As the barometer rose and fell, more of them felt sleepy.

Out of the elements of our lives, we take the symbolism of our dreams. If we see clouds in our dreams, we might have our heads in the clouds. If clouds come between us and our friends, it may symbolize clouded visions.

The appearance of the blue sky in dreams may represent creative potential or just aimless evasion. Michelangelo used the sky as a symbol of truth in his paintings. A hand coming from the sky is a sign from God.

Fog is a symbol of not having the foggiest idea of what to do. Rain in a dream can be a symbol of deliverence from a dry period of your life. If you are standing inside a house waiting for the rain to stop, it is a symbol of an obstacle in your life.

Lightning is an ancient symbol of spirit striking matter. Flashes of lightning may be symbols of insight. The insight could shatter a present relationship. Lightning can also symbolize the soul being illuminated with divine consciousness, such as St. Paul experienced on the road to Damascus.

Storm dreams occur with outbursts of emotion or desire. If your job, associates, or spouse appears in a storm dream, it may be a symbol of conflict and the possible break-up of a relationship.

All of the weather elements in our dreams are subject to our own experiences, and we alone must look to our lives to find meaning in our subconscious.

ARTHRITIS FORECASTS

As old sinners have all points, of the compass in their bones and joints, can by their pangs and aches find, all turns and changes of the wind.

—ERASMUS DARWIN

Some men do carry in their backs, prognosticating aching almanacs. Some by painful elbow, hip or knee, will shrewdly guess what weather's like to be.

—JOHN TAYLOR, "THE WATER POET"

A coming storm your shooting corns presage,
And aches will throb, your hollow tooth will rage.

—RICHARD BROME

Strength and power are located in our joints. These become deficient under the influence of the south wind; the viscous fluid in the joints becomes solidified and interferes with our moving. On the other hand too much liquidity interferes with distension. North winds make for a certain proportion so that we have strength and can stretch.

—THEOPHRASTUS

"The teletype is broken, find someone with rheumatism, so we can get the weather forecast" is an old newspaper joke. The weather sensitivity of people with arthritis is proverbial, and it is not surprising that several medical studies have been made of it.

A seven-year-old girl suffering from arthritis was asked if the weather affected her. She replied, "Yes, whenever we have a storm, my joints feel stiff and I feel sore all over; and after the storm has gone, I feel pretty well again. But I am sometimes better before the rain stops."

The first study on the relationship of pain to the weather was made by Dr. Weir Mitchell in 1872. He served in the Civil war and had done many amputations and fittings of artificial legs. Dr. Mitchell questioned 50 amputees, and found that ⅔ of them believed that they could predict storms by the increase in their pain. Pain occurred in very cold and very hot weather, but it normally occurred on a falling barometer or when there was an intense display of northern lights.

Dr. Mitchell chose one man who seemed unusually sensitive and studied him for 2 years. The patient was able to predict accurately nearly every major storm that passed near his house. He concluded that every storm has a "neurologic belt," and sensitive people in the low-pressure area will suffer until the barometer rises.

The first modern study of "arthritis weather" was done at the Mayo

178

Clinic. Patients rated themselves "better," "same," or "worse." The results were plotted in millibars of barometric pressure.

Barometric Readings (in millibars)		Worse	Same	Better
falls	−15	−4	14	5	1
slight	−3	+4	4	21	6
rises	+4	+6	1	1	11

A more detailed study was done at the Mayo Clinic to see what other factors might influence us. Most of us feel more cheerful on days of sunshine and depressed on dark days. In this study, however, sunshine had little to do with the pain of arthritis. Humidity caused a slight increase in pain, and cold also increased pain. This additional study revealed just how sensitive patients are to barometric changes.

Barometric Readings (in millibars)		Worse	Same	Better
+17	+28	0	4	0
+5	+16	8	42	9
0	+4	15	98	36
0	−3	27	79	17
−4	−11	19	41	6
−12	−20	4	3	1

Better or worse is a subjective factor, but most people with arthritis can detect a pressure difference of 7 millibars. It was found that 72% of the patients experienced more pain in a falling barometer. Only 7% felt nothing and 21% felt less pain. These results indicate that there are two kinds of arthritis.

U.S. Navy divers with arthritis have found that the pressure of deep diving relieves pain for some time afterward. The decrease in pain may be due to the fact that at higher pressures, carbon dioxide unites with calcium compounds to form soluble calcium hydrogen carbonate. As the barometer falls, we lose more carbon dioxide from our lungs, and the calcium in our joints becomes less soluble. If you are looking for a hobby that alleviates arthritis pain, why not take up scuba diving?

The theory accepted by most doctors is that the joints contain a chemical called hexosamine. As the barometer falls, less hexosamine is excreted in the urine. The chemical stiffens in cold weather, thus we have greater difficulty moving our joints. The best way to increase hexosamine excretion is to take hot baths.

Additional studies reveal that arthritics cannot tell the difference between warm or cold fronts. Osteoarthritis patients do not respond to storms, but their condition worsens as the temperature rises. Rheumatoid-arthritis sufferers are affected by humidity and generally feel worse about 12 hours before rain.

After a series of poor weekend forecasts by the English weather bureau, a leading newspaper organized the "Bunion Brigade." Before each weekend, a group of farmers, fishermen, and arthritis sufferers combined their observations for the forecast. The fishermen traditionally relied on tides, clouds, and waves. The farmers often kept seaweed, which becomes moist when the relative humidity rises because the magnesium chloride holds water.

Both group's forecasts were generally much the same for most weekends, but the Bunion Brigade had one complete miss. The official forecast was generally more detailed and usually more accurate. After 8 weeks, the final score was 61% accuracy for the Bunion Brigade and 81% accuracy for the meteorological office.

MYSTERY OF WEATHER BEHAVIOR

Mount Troldjol, also called the "rock of wonders," rises along the shores of Jorend fjord along the Norwegian coast. When the weather is about to change, columns of flame and smoke followed by thunder escape from a crack in the mountain walls.

A perpendicular rock wall 3,600 feet high at the entrance to Lyse fjord has another cave of tempests. In warm weather, when the wind hasn't blown from the southwest for several days, yellowish gray smoke comes from the caves in the rock. During strong east winds flashes of lightning shoot from the black rock, while peals of thunder echo from the cliffs.

During a mapping survey of Norway in 1855, Otto Krefting noted the mysterious phenomenon of the weather sensitive cliffs, and talked to the nearby inhabitants about it.

Whenever there is an earthquake, we humans are the last to know about it. Sensing an earthquake, dogs bark, horses are restless, birds leave the trees, cats won't come inside, and deep-sea fish come to the surface. Some of these reactions also occur in exceptionally severe storms. Why is life weather sensitive?

The weather doesn't bother adults, perhaps because they are so used to their comfortable dwellings. It does affect children more, and the earliest reports on weather sensitivity were based on observations of children. A century ago, it was noted in Denver, Colorado, that under normal humidity, children behaved properly, but under low humidity "excess flogging" went up 400%.

In 1964, a study of Wisconsin school children indicated that temperature and humidity had little to do with behavior. The study found that low pressure was associated with poor conduct. A high barometric reading brought better behavior and less physical activity, but it decreased mental activity. Students were more quiet on clear days and more restless on cloudy days. Test scores were highest during the times of greatest restlessness.

The "why" of weather behavior has been a subject of speculation for centuries. The old jewelers knew that watchsprings broke more often when there were more thunderstorms. A study of this strange mystery revealed no esoteric causes in 1922. Tiny spots of rust could be seen on the edges of the broken springs under a microscope. When springs were kept in excessively humid conditions, there was a high failure rate.

There is a considerable amount of lore about thunderstorms souring milk. Several studies have been made to determine if this is true. When milk was exposed to radioactivity, nothing happened. It has been found that bacteria grow much faster in the presence of positive ions, but that condition doesn't usually exist during storms. Yeast fermentation rises just after a cold front passes but decreases before a warm front. Sour milk was no fun, a century ago, but we have refrigerators today, and thunderstorms are no longer souring our milk.

Low-frequency sound waves have also been blamed for weather behavior. We cannot hear them, but they are shaking the air around us, and they travel far in advance of storm systems. One study shows a correlation between infrasonic waves from distant storms and absenteeism of school children and car accidents. The relationship may be extremely minor, and further studies need to be done.

We cannot hear the pulses of radio static unless we are listening to our radios. Researchers have generated intense bursts of sferics by pulsing million-volt bursts through the air. In one experiment, a bird flew near a pulse. It fell from the sky and flopped around on the ground for several minutes before flying away.

The researchers became interested with the effects of radio static on people's health. They subjected a group of rats to a pulse before having them run a maze. The rats were disoriented at first, but they recovered completely in 30 minutes. Airplane pilots have had a similar reaction after a lightning strike. Their thoughts and motor reactions are temporarily disrupted, but they recover.

There is a theory that radio static energizes cells. George Lakhovsky, the great French electronics pioneer, made special static-producing devices to energize people and cure diseases. His book: *The Secret of Life* is a classical study in a weather mystery that may be affecting everyone.

It has been found that, two days after an intense display of northern lights, more people have nervous trouble. Studies have shown that if solar flares occurred at the time of your birth, you are more likely to be emotional. Another study showed that rat mothers exposed to alternating magnetic fields gave birth to nervous offspring. There may be some truth to astrology, but it isn't in the planets, it's in the conditions that influence the weather.

The greatest correlation between our feelings and the weather is due to ions in the air. On a normal day, a cubic centimeter of air contains 1,200 positive ions and 1,000 negative ions. These negative ions are generally oxygen with an extra electron (O_{2-}) and the positive ions are carbon dioxide minus an electron (CO_{2+}).

The positive ions make it difficult for us to breath. Fatigue, stuffiness, and sore throats are more common. Temperatures flare, and judges in countries where positive winds are common have been known to be more lenient toward people.

This mystery is related to the chemistry of our brains. When we are injured or in pain, there is more 5-hydroxytryptamine (5-HT) in our bodies. This is an inflammatory chemical which increases the level of pain. Burn victims find that negative ionizers reduce the amount of pain during thundershowers, the negatively ionized air reduces levels of 5-HT. This means that we have less pain and we are less likely to have headaches. The weather affects our feelings in many ways, but this link to the recesses of our nervous system is its most interesting manifestation.

THE WINTER WEATHER GAME

And the Lord went forth down a long and ancient road and there was met by an exceeding large black cloud; and the Lord spoke thus to it: "Where goest thou?" Then spoke the cloud, "I am sent to do an injury to the poor men, to wash away the roots of his vines, and to overthrow the grapes." But the Lord spoke, "Turn back, turn back thou big black cloud, and do not wander forth to do an injury to the man but go to the wild forest and wash away the roots of the big oak tree and overthrow its leaves. Saint Peter, do thou draw thy sharp sword and cut in twain the big black cloud, that it may not go forth to do injury to the poor men."

—A SIXTEENTH-CENTURY MEDIEVAL FORMULA TO KEEP AWAY STORMS.
UNDERNEATH THE MANUSCRIPT, THE AUTHOR WROTE:
PROBATUM AN SIT ME LATET PROBET QUICUNQUE VULT
("I DON'T KNOW WHETHER IT HAS BEEN PROVEN").

The hundreds of bears in Yellowstone National Park know exactly when winter begins. When the temperature dropped to 12°F on September 15, 1965, the bears went about their normal activities.

A month later, the bears became nervous on a warm, sunny day. Late in the day, they began pacing about the entrance to their dens, but none of them entered. On November 11, all the bears headed into their dens, and a heavy snowstorm marked the real beginning of winter.

For 7 years, scientists tagged bears with radio transmitters and studied their activities. The bears were never deceived by false clues, such as cold weather and heavy snow. Each year, winter began on a different date, and the bears always knew just when this was.

We humans have played the winter weather game, but we understand it a lot less. Each fall, our newspapers and almanacs contain observations about the severity of the coming winter. They are based on the thickness of animal hair and cornhusks, the time of bird migration, and many other old weather signs.

John Burroughs, naturalist and writer, made his reputation as a winter prophet by noticing a Bohemian waxwing on the first day of December in Detroit. This bird breeds above the Arctic circle and doesn't normally appear so far south. Burroughs stated that it would be a cold winter, and he was right.

The thickly haired caterpillars known as "woolly bears" are the larvae of the tiger moth (*Isia isabella*). They feed on ragweed in the fall and then hibernate before becoming moths in the spring. Their bodies are composed of segments, which have either brown or black hair, and the width of the brown segments supposedly foretells the severity of the winter.

183

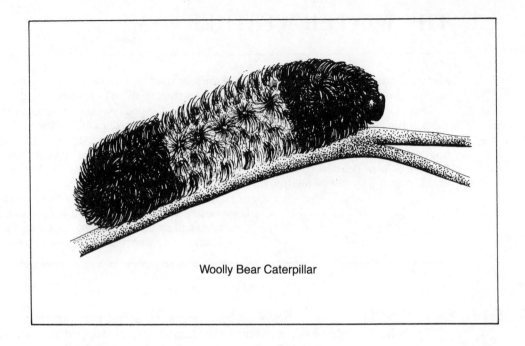

Woolly Bear Caterpillar

Dr. Charles Curran began a study of these caterpillars in 1948. During the first three years, his caterpillars had wide brown bands, indicating mild winters. These forecasts were accurate, but the caterpillars failed the next winter. Dr. Curran gave up his caterpillar study in 1955 after he found one group of caterpillars indicating a hard winter and a nearby group predicting a mild winter.

It is a little-known fact that many butterflies have summer and winter forms which are marked by different color patterns. A summer form can be changed into a winter form by putting the pupae into a freezer for a few weeks and then warming them up to hatch. It is even more interesting that a summer form changed to a winter form retains this coloration for the next few generations even if the pupae hatch in summer. Before this was known, taxonomists often classified such butterflies as two separate species. Seasonal dimorphism may have something to do with the mystery of the caterpillar's stripes, and a study might contribute to a full explanation of the mystery.

Dr. Frank Lutz accidentally discovered part of the mystery of the brown caterpillar bands in 1908. He found that a group of caterpillars raised in a wet environment were nearly all black. Raised in dry conditions, they had wide brown bands. The woolly bears were not predicting the coming winter; they were reacting to past conditions.

German immigrants brought with them the tradition of using the

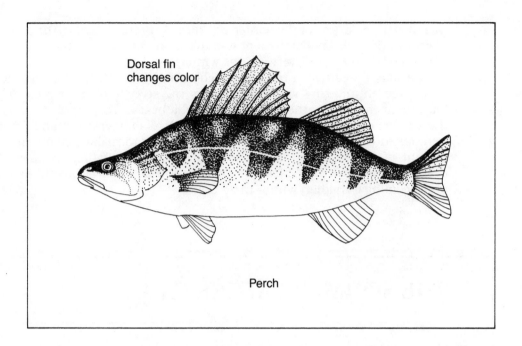

Dorsal fin
changes color

Perch

breastbone of a goose to foretell the severity of winter weather. A thin bone means a mild winter; a thick bone, a severe winter. A whiter bone means more snow, and a redder one means colder temperatures.

Every year, Chicago newspapers carry the predictions of Mathon Kyritsis. He is a commercial fisherman who bases his forecasts on the behavior of perch in Lake Michigan. If the perch head to deep water early in the fall, he says, the winter will be early and severe. If the dorsal fin is light in color, the winter will be mild; if it is dark, it will be a severe winter.

It has been found that the thickness of animal hair is a result of fall temperatures. If the fall has been cold and wet, animals will have long, thick hair. This often happened to Texas cattle shipped to Canada to graze during the summer months half a century ago. The hair length increased dramatically in the cold Canadian weather, but the hides were thinner.

It is commonly believed that the amount of nuts the squirrels put away is an indication of the coming winter's severity. But squirrels harvest only what is available to them. During the time the nut-bearing trees were in bloom in the spring, the number of flowers pollinated determined the number of nuts. If there was bad weather during those days, the squirrels have few nuts to gather in the fall.

The Indians predicted the coming winter's severity based on the height of prairie-dog mounds. They believed that prairie dogs built their mounds above the snow level so they could breathe. Another common Indian belief was that the coldness of the winter was directly proportional to the amount of rain in the fall. This statement was tested by the weather bureau, but it bore no relation to the severity of the winter.

Although weather-bureau researchers scoff at winter folklore, they discovered an amazing way of predicting the severity of the winter. They found that the fall weather of northern India bears a direct relationship to the severity of the Canadian winter. If pressure, temperature, and rainfall were taken into account, it was an accurate way to make a prediction 81 percent of the time. In 28 fall seasons in India in which temperature deviation was more than 3°F above or below normal, it paralleled the severity of the coming Canadian winter in 26 cases.

THE SPRING WEATHER GAME

Many, many welcomes, February fair maid,
Ever, as of old, solitary first thing,
Coming in the cold time,
Prophet of the gay time.
Prophet of the roses.
Many, many welcomes, February fair maid.

—THE "FAIR MAID" IS THE MILK FLOWER (GALANTHUS NIVALIS),
WHICH WAS SUPPOSED TO SHOW UP ON FEBRUARY 2.
THE AUTHOR IS UNKNOWN.

Some people think that butterflies are the most reliable sign of spring, on account of the extremely delicate nature of that insect, but in the year in which I am writing, it has been noticed that three flights of them were killed one after another by the cold weather, and that migrating birds arriving on January 27 brought hope of a spring that was soon dashed to the ground by a spell of very severe weather.

—PLINY THE ELDER, NATURAL HISTORY

Each February 2, television cameras focus on the ground hog. If the ground hog sees his shadow, there will be six more weeks of winter, and if he doesn't, there will be an early spring. The ground hog is also called the woodchuck. He emerges from hibernation at this time of year. He will have to sleep for about 6 more weeks before he can find spring food.

This curious weather observance might have been just another piece of forgotten lore if the Ground Hog Club of Punxsutawney, Pennsylvania, hadn't revived it. On February 2, 1898, 7 men from the town met on Gobbler's Knob to drink beer and eat ground hog. Their spring fun lead to this little hill being titled "the weather capital of the world." The ground hog is not related to the pig, but to the rat. He eats clover and munches on roots from the safety of its 40-foot burrow. He is good eating, and hunters often make him a target. One weatherman kept track of 15 years of the ground hog's predictions. He predicted 9 late springs, of which 4 were; and 6 early springs, of which 2 were.

In 1921, newspapers carried the story of John Willheimer's ground-hog experiment. He imported 5 of them to Monmouth, Kansas, to "settle this thing once and for all, if the ground hog really determines the length of winter." On February 2, he was waiting in the cold for his ground hogs to show their furry heads. His son finally came by and said; "Sorry, Dad, I didn't mean to do anything wrong, but there won't be any ground hogs sticking their noses out. Last fall some 'possum hunters came by, and I sacked the ground hogs and sold them for two dollars apiece as 'possums."

Groundhog

In traditional lore, it was not the ground hog that was important. Rather, it was the day that counted. February 2 was Candlemas Day, and that day was so special that any events happening on it were regarded as predictions. The lore of the day remains the same without the ground hog. This is an English verse about this special day:

> If Candlemas Day be fair and bright,
> Winter will have another flight;
> But if it be dark with clouds and rain,
> Winter is gone and will not come again.

The Nestorian Christians of Iran held a spring ceremony around Easter to determine if winter was over. A cross was put in a bowl of water and left outside the church. If there was any ice in the bowl, it would be a late spring.

At the start of each year, there were several traditional European ways of delineating the character of the following year. The character of the year's first 12 days was believed to correspond to the 12 months of the year. Some people sliced an onion into 12 pieces and sprinkled salt on them. They were put away and then examined several days later. Wet-looking slices indicated wet months.

The weather of any given month was once believed to predict that of the months ahead. For example:

> If January has never a drop, the barn will need an open prop.
>
> If in February there be no rain, it is neither good for hay nor grain.
>
> March damp and warm, will do the farmer much harm.
>
> April cold and wet, fills the barns best yet.
>
> Cold May and windy, barn filleth up finely.

The date-month items of traditional folklore have been studied against weather records. None of these old rhymes have any real value. The only ones with real truth are "all signs fail in dry weather" and "a cold, damp fall indicates a hard winter."

The reason why February 2 became such a significant date is due to Christian tradition. When the birth date of Jesus was finally put at December 25, the next significant event in his life became the presentation at the temple as described in Luke 2. The event was celebrated by the church by having a parade of lit candles to commemorate the prophecy of Simeon that Jesus would become a light to lighten the Gentiles. The date become known as Candlemas, and eventually assumed prophetic status for the year ahead.

Neither the ground hog nor traditional proverbs are of help, but there are other factors on February 2 that may be predictive. If the day is cold, the month is likely to be cold, but if it's mild, it will probably be a mild month. Some predictive surprises emerged from a study of February 2 in England in different years and the weather throughout the rest of the month.

Candlemas Day	February Mean Temperature		
	Cold	Average	Mild
Cold and easterly	6	1	1
Cold and westerly	2	0	0
Mild and easterly	0	1	0
Mild and westerly	1	6	8

The study was done in England, so it may not apply to the United States. It was suggested that the traditional rhyme be changed to read as follows:

> If Candlemas Day is cold and easterly,
> The rest of the winter will be beastly.
> If Candlemas Day is mild and westerly,
> The following weather will be besterly.

PRAYING FOR RAIN

If the officials pray for rain, and rain falls, is that why it falls? There isn't any reason for this. If they don't pray for rain, rain will still fall. When officials save the sun and the moon from being eaten during eclipses; or when they pray for rain during a drought; or if they make an important decision after divination; this is not because they think they will get the results of their wishes, they do this because it is expected of them.

—HSUN CH'HING, 300 B.C.

The idea that God sends rain in response to prayer is found in the writings of Moses: "If you walk in my statues and keep my commandments and do them; then I will give you rain in due season." If no rain fell, then people thought, that somebody had sinned, or that they hadn't kept all of the commandments.

While the Hebrews prayed to Jehovah, the Greeks prayed to Zeus. The Greek playwright Aristophanes has Socrates saying, "There is no Zeus."

Strepiades replies, "But who makes it rain?" Marcus Aurelius wrote, "The Athenians pray, 'Rain, rain, dear Zeus, upon the fields and plains of Athens.' Prayer should either not be offered at all, or else be as simple and ingenious as this."

The Bible speaks of the "early and the latter rain." The early rain came about the time of the fall equinox and generally totaled about 3.5 inches. If it did not come, it was a time for prayers. The latter rain occurred about the time of harvest during the month of Nissan, which is our March and April. It was generally about double the amount of the early rain.

The Jewish rabbis added requests for rain onto the normal Sabbath prayers when needed. The principal rain prayers were made on the day of atonement. They mentioned all the patriarchs and prophets who had anything to do with water or rain. Jehovah was asked, "In their merit, favor us with abundant rain. For a blessing and not for a curse, for life and not for death, for plenty and not for famine. Amen."

As Christianity developed in the Roman empire, there was a struggle over who should get the credit when rain fell. On one occasion, the "Thundering legion" was trapped, and the enemy planned to kill or capture them through thirst. But a tremendous thunderstorm gave them ample drinking water. Most of the soldiers gave thanks to Jupiter, but the Christians claimed that their prayers had been responsible. At this time, there was a Roman saying, "Rain falls, the Christians are to blame." Each time a rain brought needed relief, the Christian minority would claim that their prayers had brought it.

If rain is sent by God to those who have the right faith, then we would be able to decide which religion is the true one. In times of drought, we could give each group that claims to have the "only way" one day to ask God for rain, before the next group gets the chance to prove that it is the "right" group. As a control, we could also have a group of hardened gamblers who gaze at the sky while throwing dice and drinking whiskey. I doubt whether any group that claims divine authority would submit to the "rain test."

Two attempts have been made to find out if prayer has any validity. Francis Galton surveyed the health records of lawyers and clergymen. He reasoned that the preachers would be praying for divine assistance and the lawyers would be cursing their bad luck. When his statistics proved that lawyers lived just as long, he concluded that prayer has little validity.

In 1963, London doctors put prayer to a controlled test. They paired 32 people with incurable problems, such as arthritis, who would only be expected to get worse. Prayer groups flipped coins and then spent 15 minutes each day praying for each of the 16 people. There were no significant health differences between the patients who were prayed for and those who were not.

The year 1930 was marked by an exceptional drought, which led to the Dust Bowl. Thousands of farmers lost everything on the Great Plains. A lot

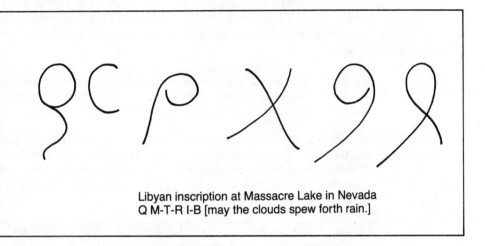

Libyan inscription at Massacre Lake in Nevada
Q M-T-R I-B [may the clouds spew forth rain.]

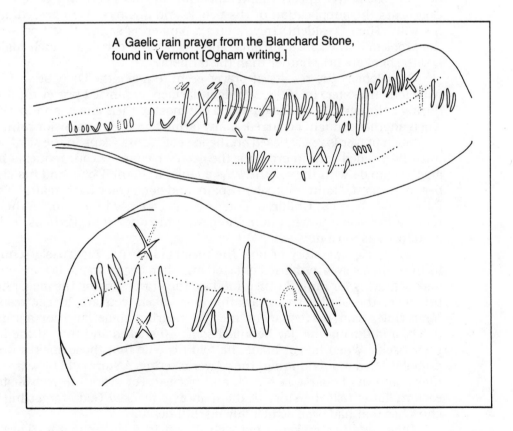

A Gaelic rain prayer from the Blanchard Stone,
found in Vermont [Ogham writing.]

of praying went on, but no rain fell. Late that year, the *Christian Century* asked 10 of the most famous preachers on their views on praying for rain. Nearly all of them said that praying for rain does not work! Here are two of their responses:

> If prayer affects the weather, meteorology ceases to be a science and becomes an article of theology. If God answered prayer for rain, then we could reasonably charge him with responsibility for drought and all other calamities.

> We might equally ask ourselves, "Does prayer affect the earthquake, the typhoon, the tornado, or the destruction wrought by lightning?" These, like a great drought, belong to those strong forces of nature that seem utterly regardless of mankind. The most difficult problem of evil is that which lies outside all human responsibility.

The head of the famous Moody Bible Institute James Grey wrote, "I do not suppose prayer affects the weather directly, but I certainly believe that God hears the supplication of His people and answers them according to his will." Then President Grey quotes a banker as saying that God solved the problem of the wheat surplus, and the price of corn was much higher. Did the bankers pray harder than the farmers?

Occasionally the results of prayer seem to make the Diety seem like a malicious trickster. In 1908, Michigan farmers organized prayer meetings during a drought. Flash floods wiped out most of their crops, and in 1923, when another drought occurred, many farmers refused to pray for rain.

Much prayer is really based on the idea of "Santa Claus in the sky," who must be pleaded with in order for the goods to come. Do our requests have anything to do with the weather? As a child, J. Frank Dobie and his sister began to chant, "Rain, rain, go to Spain, and don't come back again." Their father said sharply, "Children, don't say that, we need the rain." Although their father was a praying man, he never prayed for rain. Perhaps he had lived in Texas too long.

Dobie tells the story of how the members of the "Big Tussle" church fortified themselves with fried chicken and then began an all-day-and-night task of praying for rain. Brother Judd led the prayers, using the megaphone principle: the louder you pray, the more likely you are to get results. Meanwhile, Sam Barrow spent the day clearing some land for a turnip patch, and during the day he killed some 40 snakes and turned them on their backs. When he got home, he told his wife that those snakes would probably fetch more rain than the prayer meeting. A storm did blow up that night, and his house was struck and burned. For years the revivalists of eastern Texas told the story of the man who mocked God—forgetting the churches that had been struck during thunderstorms.

Other weather makers approached God in a different way. They assumed that they were acting for God in bringing the necessary rain. Dancing, chanting, and visualization were substituted for prayer.

The Yokut tribe of California burned grass and weeds from their fields before planting, and then they hired a rain maker. In 1870, the Indian agent at Fort Tejon promised a rain maker named Hopodno a large supply of trade goods if he would bring rain to the drought-stricken area. That night, such heavy rains fell that the agent became alarmed and asked him to stop the rain for an additional payment. The rain quickly stopped, but the extra trade goods aroused so much jealousy that the rain maker had to leave Fort Tejon.

In 1877, the rain maker's grandson (also a rain maker) was attending a gathering near the town of Porterville, California. A drunken Yokut ridiculed the young rain maker, and there was a brief fight. The angry rain maker shouted, "I'll show you!" He sprinkled water around, rolled on the ground, and threw dust into the sky while chanting. In minutes, wisps of clouds appeared, and angry thunderheads followed. Rain and hail began to pour from the skies, and everyone was scared. The chief went to the rain maker and offered him a large sum of money to stop the downpour, and it quickly stopped.

In 1935, the English writer Joan Grant went on an archaeological expedition near Bagdad, Iraq, with her husband. Day after day, the temperature rose above 100°F, and dust storms filled the air with fine, choking dust. One day, she simply couldn't stand it anymore, so she prayed intensely for hours until the sweat ran off her face. Then she ran outside naked, shouting and dancing for rain to come. The other members of the party felt that too much desert time had driven her crazy. But heavy rains fell within a radius of 4 miles from her tent that night, and puddles of water were everywhere for the next few days.

The oldest rain prayer in North America is chiseled on the Blanchard stone in Vermont. It is written in Ogham, an ancient Celtic writing using vertical lines to mark stone. It reads: "To the goddess Bianu-Mobona [earth mother]. Give thanks for rain showers by chanting for blessings and pray to Lug during Caitean [May]. Each time smoke the sacred tobacco pipe."

The famous Notre Dame football coach Frank Leahy was once asked by a reporter, "Do the pregame prayers help your team to win the games?" Leahy replied, "I think they do, but the prayers work better when the players are bigger." This seems to be true of the weather, too: The prayers are more effective when the clouds are bigger.

ANCIENT RAIN MAKING

In 1923, a terrible drought struck China. The Tashi Lama was living in exile in Beijing at the time, and as a last resort, the authorities appealed to him to do something. Under a hot, cloudless sky, some 10,000 people gathered in the central park in Beijing. The disciples lit incense, and the Lama drew a circle and prayed, "Lord of the world. I swear by Thy great name that I will not move from this place where I kneel until Thou take pity on Thy children and sendest rain." Nothing happened for several hours, and the crowd began to disperse. At 4:00 P.M., the sky began to cloud over, and a cloudburst broke out that lasted for 2 hours.

The earliest account of rain making is the Biblical struggle between Elijah and the prophets of Baal. The prophets danced and called for rain. When this didn't work, they cut themselves with knives and pleaded with Baal to send rain. Elijah made a trench, put his sacrifice in the middle, and poured 12 barrels of water over it. Fire from heaven ignited the sacrifice, and a rainstorm began.

The Biblical prophets had schools that taught secrets somewhat akin to what can be found in our books of magic. For centuries Jewish ritual fires were lit by pouring water over slaked lime, made by burning seashells. The heat thereby generated can easily ignite tinder. A few years ago, a snowfall on slaked lime set a woodpile on fire and almost burned down a house near Bend, Oregan.

Most of the ancient rain making ceremonies were imitative magic. There were dances that involved imitating clouds, flashes of lightning, and thunder. These ceremonies still continue in remote areas of the world.

Marco Polo was the first to witness the rain making magic of the East. He wrote, "If it should happen that the sky becomes cloudy and threatens rain, they ascend the roof of the palace where the grand Khan resides at the time and by the force of their incantations they prevent rain from falling and stay the tempest."

In the high plateaus of Tibet, many towns had special rain makers who were called upon in times of drought. The first Westerner to witness an actual appeal for rain was Armory de Riencourt. He entered Tibet in 1947 before the Chinese invasion destroyed the ancient ceremonies.

It was a time of drought, and the government officials decided to consult the state oracle of Lhasa. The ceremony began by burning incense and drinking buttered tea. The oracle sat on a throne as he went into a trance, with four monks holding his swaying body. The Tibetan cabinet bowed before the oracle and began to ask him questions on affairs of state. When asked if he would bring rain, he replied, "yes." Then he collapsed, and the four monks carried him out. At 11:00 P.M., thunder and heavy rains were echoing down the Lhasa valley.

The Dalai Lama had a weather-controlling monk assigned to keep his garden free from hail, which is a real problem in Tibet. In 1933, his garden was flattened, and the monk was brought in to be punished. He was told he could escape punishment by performing a miracle. He asked for a sieve and carried water in it. Thus, he escaped punishment.

The Hopi Indians grow crops in the southwestern United States on land that receives less than 10 inches of rain per year. These people have a series of ceremonies that involve appeals for rain. The Soyal festival, which involves handling rattlesnakes, is done for the purpose of bringing rain: rain is expected at the ceremonies' conclusion.

Newspapers have reported several instances in which rain dances seemed to bring rain or snow to dry areas. Three Pueblo Indians did a rain dance in San Francisco on December 12, 1960. Two hours later, rain fell—the first rain in 85 days.

The Indians of the Klamath valley of northern California have a special rain stone carved from flat-topped soapstone with 48 holes in its surface. In 1959, a road-building crew accidently uncovered it. The Huroc tribe warned local authorities that uncovering the stone would bring excessive rain. When 5 inches of rain fell in the following days, the authorities ordered that the stone be reburied.

In 1966, the stone was again dug up by a group of folklore hunters, and the next day floods washed out roads all over the area. The Indians buried the stone more deeply in a hidden place. They claimed that, in the past, deer used to jump on the stone. This would accidently bring rain, and that is why they buried it originally. They would only dig it up and hammer on it when there was a long drought.

Meteorologists think of weather in terms of air masses interacting with each other. The rain makers used imagination, and they thought in terms of a personalization of the elemental forces. Is there a place where imagination touches spirit and rain falls from the skies? Scientifically speaking, this is impossible. Many people have had unusual experiences, and they can accept psychic forces. But the display of psychic power is elusive and often dry clouds are the only result.

Sydney, Australia, needed rain desperately back in 1965. Officials brought in a band of aboriginal rain makers on September 27, and the rituals began. The next day was the hottest September day on record, and no clouds were present. Newspaper reporters crowded into an air-conditioned motel room, while the tired rain makers sipped cold beer. When asked, "What went wrong?" the aboriginal leader Robin Quartpot replied, "All dance does is to ask God to send rain. We can do no more."

MODERN RAINMAKING

The city council signed a contract yesterday with Hatfield, the moisture accelerator. He has promised to fill Morena reservoir to overflowing by December 20, 1915 for $10,000. All the councilmen are in favor of the contract except Fay, who says it's rank foolishness.

—*SAN DIEGO UNION*, DECEMBER 14, 1915

 The great inspiration for the modern generation of rain makers was the idea that major battles are followed by rain. Several dozen rain makers came across the idea that explosives, cannon booms, and gunpowder produced rain. This idea was thoroughly discredited in World War I when studies of rainfall revealed no significant increase.

The first of the great "rain makers" was Frank Melbourne. He claimed to have fled Australia after his experiments produced disastrous floods. He claimed to have produced rain on 8 successive tries and was ready to turn silver-lined clouds into silver-laden pockets.

When he began his western rain-making career, he seemingly had some success, but a string of failures turned newspaper reporters into skeptics. When he failed to bring rain, he blamed the wind for blowing his clouds away. It seems he tried hardest to bring rain on days when *Hick's Almanac* and his barometer indicated there might be a chance of rain. After several celebrated failures, he quit in 1892 and took his life two years later in a shabby Denver hotel.

Charles Hatfield was born in Fort Scott, Kansas, in 1875. He began his rain-making career in 1905 by claiming that he had been a student of meteorology for 7 years. He would set up towers with boiling vats filled with secret chemicals. He lectured widely and always said that he didn't bring rain; he merely attracted clouds, and rain followed.

Hatfield had a tremendous string of successes, and a number of failures as well. His successes generated a tremendous amount of good publicity, although the United States weather bureau sent out press releases claiming that he was a fraud.

Rain had not fallen in San Diego for months, and the city reservoir was nearly empty. In 1915, Hatfield met with the city council and offered to fill the reservoir to overflowing for $10,000. He set up his towers on the first day of January 1916, and rains began pouring down. Streets were flooded, houses washed away, and the lower Otay dam collapsed on January 27 with a considerable loss of life.

Many lawsuits were filed against him, and eventually the case went to the California Supreme Court. They ruled that rain was an act of God and

Charles Hatfield

dismissed all the suits. The San Diego city council agreed that since it was an act of God, they could refuse to pay him his $10,000.

The saga of Charles Hatfield was immortalized in Richard Nash's 1954 play, *The Rainmaker.* Saul Bellow wrote a novel about his life entitled *The Rainmaker.* A movie was made of *The Rainmaker* in 1956 starring Burt Lancaster and Katharine Hepburn. In 1973, the "Pioneers of California" erected a red granite statue of Hatfield at Lake Moreno where the floods of 1916 had occurred.

Although there are still people around who claim to know the secrets of Hatfield and other rain makers, there is little demand for their services. A Redfield, California, paper printed an interview with a C. P. Weatherford in 1958, who received the gift of making rain by "thought projection" at the age of 24. He claimed that thinking rain brings rain. When asked why California was so dry, he replied that there were too many negative thinkers in California.

The most recent and controversial rain maker was Wilhelm Reich. He was a friend of Sigmund Freud and a pioneer in psychoanalysis. He developed the theory that the universe was powered by "orgone energy." Orgone energy was a mysterious substance that Reich claimed to be able to measure and concentrate for healing and vitalization.

Reich was driven out of Europe by the rise of the Nazis. He established a center in Rangely, Maine. During a long drought in 1953, he experimented with his "cloudbuster." Reich decided that he understood the mechanics of rain and announced that he would bring rain. After pointing the device at the sky for several hours, rain fell that evening. On several occasions, farmers got together and paid him for rain when it was needed.

Two years later, he took his equipment to Arizona, where he succeeded in producing rain over a 200-mile radius. He believed that deserts were caused by an excess of "DOR" (deadly orgone radiation). If you drained off the DOR, then rain would fall and the land would be fruitful. Shortly after these experiments, he was put in prison in an action brought by the Food and Drug Administration. The charge was that he claimed to cure cancer patients with his orgone box.

Dr. Charles Kelly is the only scientist who was willing to duplicate Reich's rain-making experiments. He built a small cloudbuster and tried it on 5 weekends that the weather bureau had forecast clear skies and sunny weather. Each occasion on which he tried the device was marked by cloudy skies and some rain.

The cloudbuster originally came from Reich's observation that a grounded, hollow metal tube affected a lake's waves when it was pointed at them. He believed that he was drawing off energy and thereby affecting the waves. Scientifically speaking, there seems to be no rational reason why

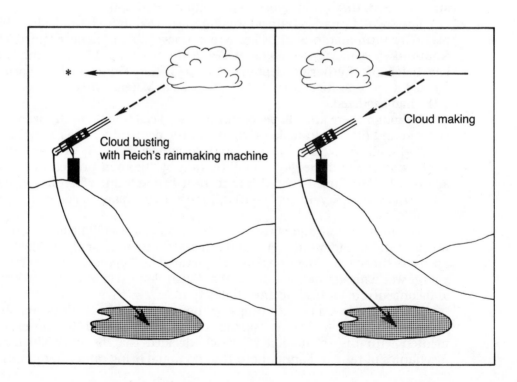

Cloud busting with Reich's rainmaking machine

Cloud making

the device should work. But practically speaking, perhaps Reich had something.

Scientific rain making began in 1947 when a General Electric scientist dropped 3 pounds of frozen carbon dioxide into a cloud. Snow began forming and turned into rain as it fell. The discovery of silver iodide lead to further cloud-seeding research. It is only effective when moisture-laden clouds are available.

Many of the ranchers of the San Luis valley of southern Colorado have contracts with the Coors brewery to produce barley. The valley has a low average rainfall, so Coors supported a cloud-seeding experiment. Cloud seeding didn't accomplish much in 1970, for it was a dry year with few clouds. The ranchers got together and demanded that rain-making experiments be stopped on the grounds that tampering with God's sky is a sin. Presumably, tampering with God's earth by irrigating and building dams isn't sinful. Coors informed them that the object of the experiments was to ensure good barley crops, and they would stop buying grain if the experiments couldn't proceed.

The experiment points out the touchy nature of weather control. For every farmer who wants rain, there's someone who wants sunshine for an equally valid reason. If there should be a possible hailstorm, in the course of cloud seeding, how can it be determined whether it is natural or artificial? And who pays?

On a local scale, we may be able to change rainfall patterns by planting forests. On a larger scale, the energy needed to shift storms and winds is beyond belief. We can artificially influence the earth's aurora with a relatively small amount of energy, and we do know that outbursts of the northern lights do change weather patterns. Real weather control may be a space-age dream, but no matter what we do, someone will always be unhappy.

THE GREAT FLOODS

The cataract of the cliff of heaven, fell blinding off the brink. As if it would wash the stars away, as suds go down a sink. The seven heavens came roaring down, for the throats of hell to drink. And Noah, he cocked his eye and said, "It looks like rain, I think." The Water has drowned the Matterhorn, as deep as a Mendip mine. But I don't care where the water goes, if it doesn't get into the wine!

—AN OLD ENGLISH ANTI-PROHIBITION BALLAD.

At times when we see buckets of water falling from the skies, we ask ourselves, "How much water is in the skies?" If all the air above us was converted into water by weight, it would form a layer 34 feet deep. If all the moisture in the air above you were to condense and fall, you would be standing in ½ an inch to 3 inches of water. The limited amount of water vapor that the air can transport is the reason why the average monthly rainfall in the eastern United States is about 3 inches. More than 2 inches of rain in a single storm is uncommon.

If a square mile of land is to receive more than an inch of rain, then the air over several square miles must be made to condense and give up its moisture. A thunderstorm can do this by lifting and cooling the air from a much wider area. By forcing air upwards at speeds of perhaps 30 miles per hour, moisture is frozen out in much the same way as it is inside a refrigerator-freezer.

On the mountain slopes of Hawaii, more than 400 inches of rain falls. The steady trade winds blow mile after mile of wet air against the sloping land. Lifting the air by 5,000 feet cools it below the dew point, and heavy rains are the result. The other sides of these mountains are rocky deserts. The great Himalayas form such a high barrier that perhaps ¾ of all the moisture in the monsoon winds is wrung out over the country of India. Cherrapunju, India, holds the world's record for rainfall: 88 feet of rain in a single year.

The record rainfall for a short duration is 12 inches in 42 minutes at Holt, Missouri, in 1947. If we accept the Biblical flood as a statement of fact, then more than 10 times this amount fell continuously for 40 days.

The unofficial record is probably held by Sand Canyon in California. A thunderstorm formed over the Mojave desert and, at the same time, another storm formed over the crest of the Rocky Mountains. The lower storm was sucked into the upper storm, and many feet of water may have fallen from the sky in a short duration. A 70-foot wall of water swept down the canyon and stripped everything, leaving only barren rock.

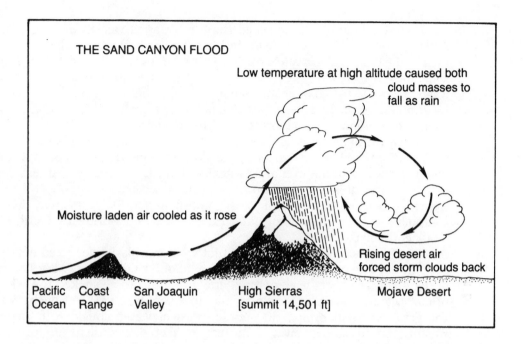

THE SAND CANYON FLOOD

Low temperature at high altitude caused both cloud masses to fall as rain

Moisture laden air cooled as it rose

Rising desert air forced storm clouds back

| Pacific Ocean | Coast Range | San Joaquin Valley | High Sierras [summit 14,501 ft] | Mojave Desert |

Each flood brings down silt into lakes or the ocean, which forms a layer known as a *varve.* The entire history of a lake can be unraveled in the same manner as a tree's rings. A comparison in varves in distant lakes builds up a picture of area-wide flooding.

The records of past floods can be found by studying *slack water deposits.* As a river floods, slack water carries debris into the mouths of slower-moving feeder streams downriver. Pits are dug near the mouth of these streams, and they show a series of layers of silt and wood. These layers are dated with radioactive carbon. This method revealed that the South Texas rivers have had one major flood every 2,000 years. Knowing the frequency of major floods helps in planning the construction of houses, highways, and bridges. If major floods were expected every 50 years, it would be important to provide a wide safety margin for buildings.

Geologists and archaeologists have uncovered many floods through studying the earth's strata. The cities of Mohenjo Daro and Harappa in India were abandoned after several floods around 1500 B.C. There is 3 to 4 feet of flood-deposited sand near Lothal, and 38 feet of alluvial gravel under Budh-Takkur in Sind, India.

When earthquakes occur, massive landslides block the rivers flowing out of the Himalayan mountains. In 1857, a landslide at Sarat, Pakistan, blocked the Hunza river for 6 months. When the dam broke, the river rose 55 feet hundreds of miles downstream.

In 1893, a landslide dammed the Bireh Ganga valley, forming a dam 900 feet high and 2 miles wide at the base. The British authorities stationed men at the dam and ran a telegraph wire to the towns. On August 25, 1884, water began to trickle over the dam, and the next day 400 feet of water washed out of the lake. Many small villages and the town of Srinagar, Kashmir, were washed away, but loss of life was minimal. In an earlier era, such a flood might have generated a mythological story.

An earthquake-generated landslide was the cause of Noah's flood. This probably occurred in the canyons of the Euphrates River near Mount Ararat. When the water broke over the landslide dam, it created the largest flood that the Euphrates has ever known. The prehistoric stories about this flood were incorporated into the Epic of Gilgamesh, and 2,000 years later, Hebrew compilers took the flood story and turned it into a worldwide event with Noah as the hero.

The greatest flood in the recent history of our planet occurred near the present-day site of Spokane, Washington, about 13,000 years ago. The last of the great ice sheets was melting, and a tongue of ice blocked the release of melt water. Missoula, Montana, was in a vast lake 950 feet deep. When the great ice dam broke, 500 cubic miles of water spilled into eastern Washington. It ripped away hundreds of feet of sediment to form the Columbia River gorge. Portland, Oregon, had 400 feet of water flowing over it for a few days.

The terrain southwest of Spokane looks like ripple marks on a beach when viewed from an airplane. When the pioneer geologist Harlen Bretz investigated this area in the 1920s, he saw it as the remainder of a great flood. His coworkers thought this explanation was ridiculous. His views were fully confirmed in 1965 when an international group of geologists examined the entire area. They found raised beaches on Montana mountains, and a canyon with a dry waterfall.

When satellites sent back the first pictures from Mars to Earth, a similar pattern of "channeled scablands" was noted. This lead to speculations that Mars once had an ice age, which suddenly melted.

The greatest floods of the eastern and southern United States are a result of hurricanes. Another weather pattern that produces flooding is a combination of high-pressure centers over the Great Plains and the Atlantic. Cold Canadian air flows between the centers, and moist air is pushed up to great heights.

The freak ocean temperature pattern known as *El Niño* can alter the rainfall pattern on the west coast. During the months of December 1861 and January 1862, the California coast received about 50 inches of rain. These rains were so heavy that settlers in Oregon and Washington recorded that their cabins leaked. There were so few people around that the effects were relatively minor. Rains of this magnitude could make an earthquake seem minor.

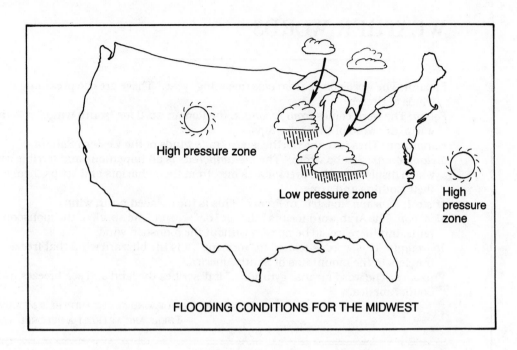

FLOODING CONDITIONS FOR THE MIDWEST

   Much of the earth's population lives on river deltas because the land is fertile and water transportation is available. Dams and dikes produce a sense of security, but any major alteration of weather patterns could threaten the security of millions of people.

WEATHER WORDS.

Etesian: The word comes from *etos*, meaning "year." These are the prevailing winds of Greece.

Foehn: The word comes from *favonius*, the Roman word for "south wind." It is the warm, dry wind of the Swiss Alps.

Harmattan: These are the northeastern trade winds of the Western Sahara.

Melamboreas: Strabo wrote, "The *melamboreas* is an impetuous and terrible wind which displaces rocks, precipitates men from their chariots and strips them of their clothing and arms."

Mistral: It means "master" in French. This is the dreaded north wind.

Monsoon: The Arab word means "change" or "season." We speak of the monsoon rains, but there would be no rain without the monsoon wind.

Tourments: This word is related to "torment." It is the blizzard wind that freezes travelers in the mountains of South America.

Vrazones: The word means "gyrations." It describes the land and sea breezes in South America.

—A HANDFUL OF WIND NAMES AND THEIR MEANINGS.
THERE ARE AT LEAST A HUNDRED MORE.

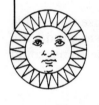

The descriptiveness of words is the measure of the ability to be understood. When we consider that the Eskimos have more than 30 words for snow, and that those who herd caribou have 30 words for these animals, we realize how poor our language is when it comes to things that we do not normally consider.

The word *weather* comes from words spoken by an ancient people more than 5,000 years ago. We don't know a lot about these people, but we know that they scattered parts of their language from Northern India to Iceland. The word *weather* is derived from *we*, meaning "wind," and *vydra*, meaning "storm." In the old Sanskrit language of India, they used the word *akasavidya*, meaning "knowledge of the air," as a word for meteorology.

Weather has about a hundred cognates such as *weather-dogs*, which we now call *sun dogs*. The *weather-head* is the secondary rainbow. *Weather-sick* is an old term for "cabin fever." *Weatherbeaten* has the obvious meaning of being aged from exposure to the weather.

The science of weather study is known as *meteorology*. It was probably chosen from Aristotle's *Meteorologikon*, which was spelled *Meteorologica* in the Latin edition. The root properly means "meteors," so it does not mean what we choose it to mean. The direction of meteors was once believed to be an indicator of future weather, so the association is not improbable.

The study of water flow is known as *hydrodynamics,* and we would naturally expect the study of air flow to be *aerodynamics.* The aircraft designers appropriated that word, so meteorologists had to settle for something less dynamic. It has been proposed that a weatherman is properly an *aeronomer* and his science is *aerology.* This proposal was made 40 years ago, but few meteorologists have seen fit to change the language.

The names for the seasons are also derived from the Indo-Germanic language. Winter derives from *wed,* meaning "to be wet," which is an indication that its ancient origin was not in Europe. Summer comes from *sam-year,* meaning "half-year." Spring and fall are derived from the English verbs "to spring up" and "to fall down."

Three thousand years ago, the women of northern Europe used the word *stirm* when they mixed cooking items together. Eventually the word separated into the words *stir* and *storm.* The word *storm* properly means "the stirring up of the skies."

Captain Piddington first used the word *cyclone* in his *Handbook for Sailors* in 1844. The term means "coils of a snake." *Typhoon* refers to exactly the same type of storm, only in the Pacific Ocean. It is a mispronunciation of a Chinese term, *ta-feng,* meaning "violent winds."

It was once believed that the Blizzard family of Buckinghamshire, England, had migrated to the west and given their name to northern storms. Other philologists believed that it was an uncommon English word, once used to describe a knock-down fight, with a "blizzard" of blows.

Blizzard first appeared in the *Northern Vindicator* of Esterville, Iowa, in 1870. Many of the early settlers in this area were from Germany, and, when witnessing the severe winter storms, would use the German expression *Der Sturm kommt blitzartig,"* meaning "the storm comes lightning-like." The transition from *blitzartig* to *blizzard* was a natural language progression.

The word *smog* was first used in France in 1905, but it went no further. The word was recreated when an inebriated reporter for the *Harold Examiner* of Los Angeles wrote a story and mixed together the words *smoke* and *fog.* The editor tightened the mixture into the word *smog* for the headline.

The Greeks used to call a spell of warm, early winter weather *halcyon days.* This was the time that the *halcyon,* a mythical bird, built its nest on the sea and hatched its young. A warm spell of December weather does not always occur worldwide, at the same time, but it is a common feature of Greek Decembers.

It is commonly believed that the 7-day week came from a rough estimation of the quarters of the moon. The Sumerian and Akkadian languages of the ancient Middle East had the same word for *day* and *wind.* These people believed in the 7 winds of heaven, and they had 7 flights of stairs up the ziggurats. The mystical idea of 7 winds became the week.

Nearly all tribes and countries once used a calendar with several weather months. The ancient Egyptian calendar began on September 11, when the star Sirius was on the horizon, and this was celebrated by the Feast of Nayroz. The months were *Toot, Baba, Hatoor, Kiyahki, Tooba, Amsheer, Baramhat, Barmouda, Bashan, Baoona, Abeeb,* and *Misra. Tooba* represented the god of rains, *Amsheer* was the god of storms, and *Bashan* was the god of sunshine.

Most of the North American tribes counted months by moons. Since there are about 12.4 moons per year, it was difficult to decide what moon one was really in. The moon months were often named for weather elements, such as snow, frost, or winds, or were named after the crops and animals.

The English calendar that we use is derived from the Roman names for the months. April, May and June are the goddesses and January and March are the gods. Julius and Augustus Caesar had their names immortalized in July and August. The months of September, October, November and December represent the numerals seven through ten. It is surprising to find that there is no weather association with the calendar months when weather is such a major feature of our life.

INDIAN SUMMER

Now summer wanes into the Fall,
The heat and strife abate.
And coolness settles over all,
The sun lags not so late.
Blue haze obscures the farthest hill,
And thoughts like cattle hug the fold.
The peaceful air is bright and still,
Who minds this growing old?

—AUTHOR UNKNOWN

The one uniquely American contribution to weather language is the term *Indian summer*. It evokes memories of warm, hazy fall days with frosty nights.

A true Indian summer rarely exists in Europe. It is sometimes called "fool's summer" or "old woman's summer," perhaps in reference to its short and unpredictable duration. Faithful churchgoers in Europe took to naming the spell of warm weather after the birthday of the saint on which it occurred. Hence, there is Saint Teresa's summer, Saint Martin's summer, Saint Bridget's summer, Saint Michael's summer, and Saint Luke's "little" summer.

The hazy appearance of the fall days is produced by frost. When water freezes inside tree leaves, it cracks the cells. The volatile hydrocarbon compounds are evaporated by the heat of the sun and the wind, and the skies have a bluish haze.

The Indians told stories about this haze. In 1861, Peter Jones, an Indian writer, recorded the following legend:

> Nanahbozhoo always sleeps during the winter; but previous to his falling asleep, he fills his great pipe, and smokes for several days, and that is the smoke arising from the mouth and pipe of Nanahbozhoo which produces what is called "Indian summer."

In the "Song of Hiawatha," Henry Wadsworth Longfellow adopted a story from the Onondaga Indians:

> From his pipe the smoke ascending,
> Filled the sky with haze and vapor,
> Filled the air with dreamy softness,
> Gave a twinkle to the water,
> Touched the rugged hills with smoothness,
> Brought the tender Indian Summer
> To the melancholy North land,
> In the dreamy Moon of Snow-shoes.

Reverend James Freeman gave us another interesting insight into aboriginal beliefs in an early letter:

> The Southwest is the pleasantest wind, which blows in New England. In the month of October, in particular after the frosts, which commonly take place at the end of September, it frequently produces 2 or 3 weeks of fair weather in which the air is perfectly transparent and the clouds which float in a sky of purest azure adorned with brilliant colours.
>
> This charming season is called the Indian Summer, a name which is derived from the natives, who believe that it is caused by a wind, which comes immediately from the court of their great and benevolent God Cautantowwit, or the Southwest God, the God who is superior to all other beings who sends them blessings which they enjoy and to whom the souls of their fathers go after their decrease.

There was a widespread belief among the early settlers that cold weather and storms came around the autumnal equinox, which occur on or about

September 23. These brief storms were occasionally referred to as the "squaw winter" or the "half winter." After they had passed, the true Indian summer began.

Some early writers placed a definite beginning and ending time for the real Indian summer. Most writers referred to it as a type of weather. Over the years, Henry David Thoreau noted in his diaries that Indian summer weather occurred anywhere from September 27 to December 13.

The earliest mention of the phrase occurs in a French letter dated 1778:

> Sometimes the rain is followed by an interval of calm and warmth which is called the Indian summer, its characteristics are a tranquil atmosphere and general smokiness. Up to this time the approaches of winter are doubtful, it arrives about the middle of November, although snows and brief freezes often occur long before that date.

The earliest mention of the English term is in General Harmar's return report, written after his Cincinnati soldiers had destroyed the Maumee Indian villages near Fort Wayne, Indiana:

> Thursday, October 21 [1790]. Fine weather—Indian summer. Having completed the destruction of the Maumee towns, as they are called, we took up our line of march this morning from the ruins of Chillicothe for Fort Washington. Marched about 8 miles.

The term spread to New England by 1798, to New York by 1809, and to Canada by 1821. Many of the great writers used it as a popular figure of

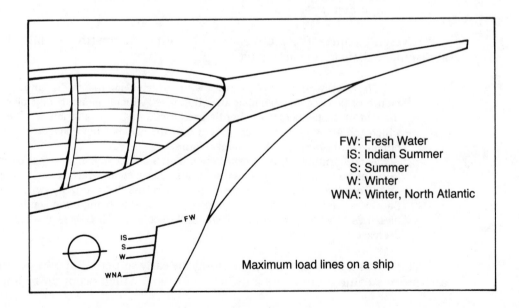

FW: Fresh Water
IS: Indian Summer
S: Summer
W: Winter
WNA: Winter, North Atlantic

Maximum load lines on a ship

speech to describe the warm, hazy fall days before the winter. It was also widely believed that the haziness came from the Indians burning the leaves and brush in preparation for next year's corn fields.

The term "Indian summer" evoked a completely different meaning among sailors. Wealthy nineteenth-century ship owners used old ships for carrying cargo. When the ship became battered and leaky, the owners carried more insurance, and if it went down in a storm, they collected. Many ships were undermanned, overloaded, and overinsured.

Samuel Plimsoll entered the English Parliament as a liberal in 1873. He began a crusade against dishonest shipping pratices, which resulted in a book of scandals called *Our Seamen—An Appeal.* It describes ship owners as greedy, wealthy leeches, and ends with the plea, "Help the poor sailors for the love of God!"

As a result of this crusade, Parliament passed laws to protect sailors by putting load lines on the sides of ships. Fresh water is lighter than salt water, so more cargo can be carried on the ocean. Cold water is heavier than warm water, so more cargo can be carried in the winter months. The extensive trade with the British East India Company lead to loading marks being placed on the front of the ship. Every longshoreman knew that the "I.S." at the front of the ship meant "Indian Summer," and some believe that this became the term "Indian summer."

MAKING MONEY ON THE WEATHER

"Well, how is the dry weather serving you?"
"How has the rain been over in your section?"
"When you reckon it's going to rain?"
"How's the crops over your way?"
"It's getting awful dry, isn't it?"

—THESE ARE THE FIVE QUESTIONS THAT ALL FARMERS FROM
TEXAS TO ARIZONA ASK EACH OTHER WHEN THEY MEET,
ACCORDING TO J. FRANK DOBIE.

 The ancient Greeks told the story of how the "laughing philosopher," Democritus the Milesian, made his money. He foresaw a shortage of olives and went about buying them all up. When the shortage occurred, he gave them to the sellers, saying, "You can see now that a philosopher can get rich whenever he pleases."

Day after day, we hear the announcer say: "The Dow-Jones Industrials are up eight points today, investors are confident over the new policies," or "Stocks are down ten points today, investors are nervous about interest rates." Do the collective pool of investors really think this, or do they simply wake up in the morning saying to themselves, "Great day, I'll buy one thousand shares of XYZ today." Then there are the dreary days when they wake up thinking, "Today I'm selling my shares of XYZ."

Las Vegas is only a minor gambling center when compared to Wall Street. But we consider it more sinful to gamble in Vegas, because almost everyone loses money. On Wall Street, smart people make money, but occasionally they, too, lose their shirts.

There are theories about how each aspect of life influences the stock market. Adam Smith described one of the more interesting ones. On days when the wind is just right, it blows the odor from a little popcorn stand along the street. People feel good, they buy stocks, and the price goes up.

The earliest investing theories were based on sunspots. In 1884, the British economist Stanley Jevons published his *Investigations in Currency and Finance.* He thought he had found a 10.4-year cycle corresponding to sunspots. It is now known that his investigations were incomplete and that the actual cycle is somewhat longer. When a whole century of Wall Street prices are compared to sunspots, there is some correlation.

Henry Moor was the next economist to try linking cosmic cycles to the price of stocks. He believed in an 8-year crop cycle, which was triggered by 11 rotations of the planet Venus. He felt that rainfall determined agriculture, which in turn determined the prices of stocks.

Perhaps the most interesting cosmic cycle to affect the weather was discovered by Robert Edwards. He found that low points in stock prices occurred at the new and full moons. High points tended to occur at the first and third quarters. The difference between these high and low points is only about 1%. If you didn't have to pay a broker's commission, in theory you could make 24% per year by following the moon.

The fastest way of making a lot of money in the market is through futures trading, and this is directly dependent upon the weather. Companies making candy bars want to guarantee themselves a reasonably priced supply of chocolate. They are willing to pay a little more for futures, for they are really buying insurance. If the harvest fails and the price doubles, they won't have to change the prices of their candy bars.

Airline pilots study weather patterns as part of their business. Several years ago, a group of pilots noticed that weather patterns made a frost likely in Florida in a few days. They bought $2,500 worth of orange-juice concentrate contracts. When the frost occurred, they sold out several weeks later for $9,500.

There is another way of gambling on the weather on Wall Street. It is claimed that it's best to sell your hot-weather stocks in the spring and buy them in the fall. Hot-weather stocks include bathing suits, air conditioners, and soft drinks. It seems that weather stocks normally tend to rise in the spring and drop in the fall.

A small company in Chicago called Sales Productivity Service was formed to exploit weather-buying patterns. They made up 5-day advertising forecasts because long-range accuracy tends to go bad quickly. In order to understand their ideas do some store observing. Have you ever been to a large store that was either crowded with people or virtually empty and wondered why? The "let's go shopping" feeling is partially triggered by weather.

Sales Productivity Service uses a "Bioclimatic Index" to determine "favorable buying days." By mailing circulars on favorable buying days, the sales of a product might double. Car salesmen report especially enthusiastic responses to advertising on favorable days. Much of this information is practical wisdom, depending upon the product. Obviously, you would mail out circulars for home insulation to coincide with the first cold weather in the fall.

It is difficult to relate the weather to the buying of stocks, because many kinds of weather are present throughout the country. Political decisions are taking place, which distort weather influence. Edgar Smith did a study of weather effects on the market. Rainfall and gray skies depressed the index; it rose with a rise in temperature. He worked out a complex formula to determine stock prices based on weather studies. Similar results have been obtained in Japan.

Weather is a constant factor in our thinking and actions, and it influences us in little ways. Our awareness of its influence can contribute to a happier life, and perhaps even to making a little extra money.

THE WEATHER KITE

When the glass falls low, prepare for a blow;
When it rises high, let all your kites fly.

The charge coming down the kite wire rendered it incandescent and made it appear slightly larger than 1 centimeter in diameter. At the reel a cannon-like report was heard, and melted pieces of wire were scattered in every direction, liberally spraying the men on duty. None of the men was injured, although the one operating the kite reel received a slight shock. Those outside the reel house stated that the building had the appearance of being in flames. Considerable heat and a dazzling white glare accompanied the phenomenon. The vaporized wire left a rocket-like trail of yellowish-brown smoke which remained visible for 15 minutes throughout the entire length of the line.

—*MONTHLY WEATHER REVIEW SUPPLEMENT,* No. 11:5, 1918

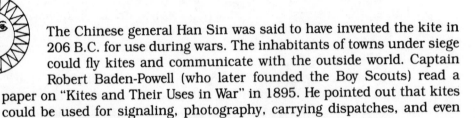

The Chinese general Han Sin was said to have invented the kite in 206 B.C. for use during wars. The inhabitants of towns under siege could fly kites and communicate with the outside world. Captain Robert Baden-Powell (who later founded the Boy Scouts) read a paper on "Kites and Their Uses in War" in 1895. He pointed out that kites could be used for signaling, photography, carrying dispatches, and even lifting observers.

The weather observers had a different purpose for flying kites. To date, the only way to find out about the upper atmosphere was to climb mountains. By flying kites or releasing balloons, they could obtain information immediately and easily.

The first person to use this method of information was Dr. Alexander Wilson of Glasgow, Scotland. He attached thermometers to the second of 2 large kites. Since they were not self-recording, he reeled in the kites and took the readings. These were the first measurements of the *lapse rate*, the rate of temperature drop with increase in altitude. It is normally about 1°F for every 300 feet of altitude.

Benjamin Franklin was the next kite flyer to investigate weather phenomena. He experimented with static electricity generated by friction on glass or sulphur globes. The electrical charges were stored in Leyden jars. He began to wonder if lightning was simply a very powerful electrical spark.

In 1752, he wrote a letter to Peter Collinson telling how he had discovered that static charges were the same as lightning. He flew a kite in a thunderstorm, and a spark traveled down the damp silk thread and jumped to a metal key. The practical result of this experiment was the lightning rod.

Franklin's account aroused a great deal of interest in Europe. The year 1752 was cold and damp, and there were few thunderstorms. Soon other scientists confirmed that kites could draw electricity from clouds. The fun continued until a Russian scientist was killed by lightning. European builders immediately adopted Franklin's idea of lightning rods, and damage became less common.

William Espy began flying weather kites around 1835. He wanted to confirm the theory that clouds were the result of a temperature drop that brought moisture to the dew point. He flew kites into the bases of clouds and calculated their height with a sextant. He found that the calculations of temperature drop needed to form clouds were accurate. He also noted the strong updrafts at the base of clouds.

In 1836, James Swain used a kite to carry an electrometer into the air to measure the "electrical density." In 1911, this same experiment lead to the discovery of cosmic rays. At higher altitudes, the leaves of an electrometer quickly collapse because of ionization reduces the charge that keeps them apart.

In 1882, Douglas Archibald used kites to carry aloft anemometers to study wind speed changes with altitude. Surface winds are slower and more gusty because of waves created by passing over objects. The speed and direction of wind changes with altitude.

In 1890, the United States Weather Bureau began a kite-flying program to take regular measurements of the atmosphere. They began studying kite designs to find something that could carry instruments aloft on light winds. The familiar birdlike "malay kite" was used until a better one was found.

It was found that the best design was the "Hargrave kite." Lawrence Hargrave was an Australian scientist who nearly beat the Wright brothers to the invention of the airplane. In 1890, he powered his kites with compressed air, and by 1892 he had a small steam-engine flying machine. He made very large kites, in which he was able to ride and make observations in high winds.

With good kites, the weather bureau was able to find out most of the facts that we now know about lower elevations. By 1896, a string of 9 kites lifted instruments to 8,740 feet. A surveyor's transit was used to determine the angle of elevation and accurately measure the height.

By using kites, they found that clouds were warmer and dryer than the surrounding air. Condensation of vapor gives up heat energy, which can form a self-perpetuating system. The kite-borne instruments determined that cold and warm waves could be detected at upper elevations earlier than they could be on ground level.

The kite flyers discovered strange facts, which were later of important use in flying airplanes. All winds turn to the right at higher elevations, so a south wind at ground level will be a westerly wind 3 miles up. A southerly wind may shift by 90°, but a northerly wind will shift by a lesser amount. At

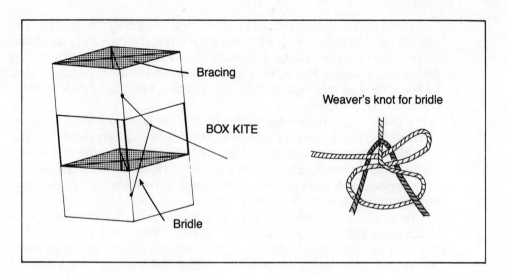

Bracing

BOX KITE

Bridle

Weaver's knot for bridle

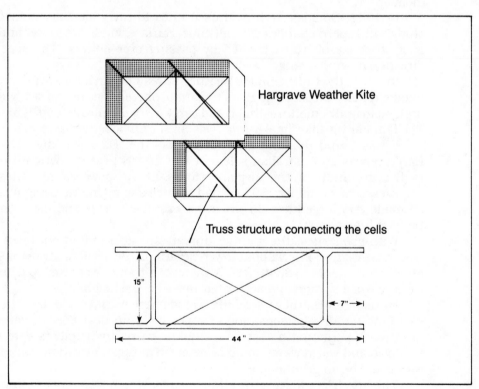

Hargrave Weather Kite

Truss structure connecting the cells

15"

7"

44"

2,000 feet above the ground, wind speeds are generally double that of ground level.

The present altitude record for kite flying is reported to be 33,000 feet. Federal Aviation Authority officials were concerned about aircraft at nearby O'Hare Airport, and they ordered the kite string cut. It would have taken hours to reel in eight miles of kite string.

In 1963, Domina Jalbert developed a flying-wing kite in which air is rammed into the leading edge. This kite was adopted by the weather bureau for carrying instrument packages upward, for it can carry heavier loads and fly higher at a more vertical angle. This kite has largely replaced the traditional canopy parachute, because it has the advantage of being steerable.

Since there are few better ways of combining science and fun, I have included the plans for the official (1895) weather bureau kite. The only change that we suggest is to add gussets, which are thin, triangular pieces of plywood glued to the corners to provide cross-bracing. This kite will fly in light winds and provide appreciation for the efforts of our weather pioneers.

BIBLIOGRAPHY

AMSB—American Meteorological Society Bulletin
MM—Meteorological Magazine
MWR—Monthly Weather Review
QJRMS—Quarterly Journal of the Royal Meteorological Society
SA—Scientific American
SMM—Symond's Meteorological Magazine
*Note: All titles have been translated, and some have been shortened.

FAMOUS WEATHER ANECTDOTES
American Forests 84:85, 1929 "Red Signs and White Science" C. E. Bosooth
AMSB 47:975, 1966 "Thomas Jefferson and the American Climate" D. M. Ludlum
Hobbies 62: Aug/114, 1957 "Indians Ability to Predict Storms Baffles Weathermen"
Morrison, Samuel E. *Admiral of the Ocean Sea.* Boston: Little, Brown and Co., 1944.

NOAH'S FLOOD
Antiquity 54:34, 1980 "The Date of Noah's Ark" R. E. Taylor, R. Berger Geographical Magazine
    38:390, 1955 "The Tower of Babel; Fact and Fantasy" H. Minkowski
Iraq 28:52, 1966 "Noah's Flood Reconsidered" M. E. Mallowan
Allen, Don C. *The Legend of Noah.* Urbana, Ill.: University of Illinois Press, 1949.

ANCIENT WEATHER
AMSB 53:634, 1972 "Aristotle and His Meteorologica" H. Frisinger
    Greece and Rome 20:26, 49, 1951 "Weather Signs in Virgil" L. S. Jermyn
Charles, Robert H. *The Book of Enoch.* London: Society for Promoting Christian Knowledge,
    1917.

THE TOWER OF THE WINDS
American Journal of Archaeology 47:291, 1943 "The Tower of the Winds and the Roman
    Market Place" H. S. Robinson
National Geographic Magazine 131: April/568, 1967 "The Tower of the Winds" D. J. DeSolla
    Price

SECRETS OF THE BAROMETER
Marine Observer 44:79, 1974 "Unorthodox Meteorology" J. Smit
QJRMS 12:11, 1886 "Brief Historical Account of the Barometer" W. Ellis
Middleton, W. E. K. *The History of the Barometer.* Baltimore, Md.: Johns Hopkins Press, 1964.

NATURAL BAROMETERS
Nature 69:7, 104, 127 1903 "Weather Changes and the Appearance of Scum on Ponds" H. R.
    Mill, W. Ramsden, F. J. Hillig
SMM 16:155, 1881 "On the Influence of Barometric Pressure on the Discharge of Water from
    Springs" B. Latham
SMM 21:57, 1886 "Barometric Wells" J. S. Harding

THE THERMOMETER
Horticulture 6:51, 1928 "Rhodendrons as Living Thermometers" G. G. Nearing
Horticulture 14:132, 1936 "A Living Thermometer" G. W. Phillips
MWR 28:551, 1900 "The Evolution of the Thermometer" H. C. Bolton
Natural History 58:256, 285, 1949 "Insect Thermometers" C. Hallenbeck
Science 119:442, 1954 "Temperature Dependence of Rattling Frequency in the Rattlesnake"
    L. E. Chadwick, H. Rahn

NATURAL HYDROMETERS

Annales of Applied Biology 32:78, 1945 "The Use of Cobalt Salts as Indicators of Humidity and Moisture" M. E. Solomon

Mentor 14: Aug/59, 1926 "Weather Toys" C. F. Talman

Proceedings of the Linnean Society of New South Wales 8:462, 1883 "The Barometro Auraucano from the Chiloe Islands" N. DeMiklouho-Maclay

QJRMS 68:247, 1942 "The Early History of Hydrometry and the Controversy Between De Saussure and De Luc" W. E. K. Middleton

THE FIRST FORECAST

MWR 51:1, 1923 "History of Radio in Relation to the Work of the Weather Bureau" E. B. Calvert

Nineteenth Century 101:557 and 102:93, 1928 "The Development of Weather Forecasting" G. C. Simpson

Mellersh, Harold E. *Fitzroy of the Beagle.* London: Hart-Davis, 1968.

WEATHER IN THE BATHTUB

AMSB 47:401, 1966 "On Coriolis and the Deflective Force" C. L. Jordan

American Scientist 71:353, 566, 1983 "Notes on the Bathtub Vortex . . ." M. Sibulkin

New Scientist 17:302, 1963 "Whirlpools, Vortices and Bathtubs" C. N. Andrade

THE WAYS OF THE WINDS

Classical Weekly 24:11, 18, 25, 1930 "Greek and Roman Weather Lore of the Winds" E. S. McCartney

MWR 47:730, 1919 "The Relation of Wind Directon to Subsequent Precipitation in Central Ohio" H. H. Martin

MWR 49:507, 1921 "History of the Theories of the Winds from Earliest Times to the Beginning of the Seventeenth Century" E. W. Woolard

Proceedings of the British Academy 6:179, 1913 "The Rose of the Winds" S. P. Thompson

THE SOUND OF THE WEATHER

Journal of the Acoustical Society of America 5:112, 1933 "The Absorption of Sound in Air, in Oxygen, in Nitrogen . . ." V. O. Knudson

QJRMS 52:351, 1926 "Meteorological Conditions and Sound Transmission" E. S. Player

SA 124:47, 1921 "Telegraph Wires as Weather Prophets"

Washington Academy of Science; Journal 13:49, 1923 "The Murmur of the Forest and the Roar of the Mountains" W. J. Humphries

THE SMELL OF RAIN

Acta Geochemica et Cosmochemica 30:869, 1966 "Genesis of Petrichor . . ." I. J. Bear, R. G. Thomas

Applied Microbiology 13:935, 1965 "Geosmin; An Earthy-smelling Substance Isolated from Actinomycetes" N. N. Gerber, H. A. Lechevalier

Australian Journal of Chemistry 18:915, 1965 "Fatty Acids from Exposed Rock Surfaces" I. J. Bear, Z. H. Kranz

Journal of Organic Chemistry 33:2593, 1968 "The Synthesis of Geosmin and the other 1, 10 Dimethyl-9-decanol isomers" J. A. Marshall, A. R. Hochsteler

CLOUD PREDICTIONS

La Meteorology 42:187, 1942 "The Meteorological Folklore of Clouds" L. Dufour

Philosophical Magazine 16:97, 344, 1803 "On the Modification of Clouds" L. Howard

QJRMS 51:191, 1925 "Clouds and Forecasting Weather" E. V. Everdinger Spectrum 105:5, 1973 "Luke Howard—The Man Who Named the Clouds" A. G. Thoday

PROPHETIC SKIES

Ciel et Terre 6:337, 1885 "Indication Colors from Star Twinkling and Atmospheric Variation" C. M. Montigny

MM 61:15, 1926 "A Red Sky at Night" S. Russell

MWR 48:511, 1920 "The Relation of Telescopic Definition to Cold Waves" W. H. Pickering

MWR 49:191, 1921 "Fire Colored Sunsets as a Valuable Clue to the Existance of a Tropical Storm" R. M. Dole

THE RISING OF THE MOUNTAINS
Science 192:1300, 1976 "The Arctic Mirage and the Early North Atlantic" H. L. Sawatzky, W. H. Lehn
SA 234: Jan/102, 1976 "Mirages" A. B. Fraser, W. H. Mach
Stefansson, Vilhjalmur. *Ultima Thule: Further Mysteries of the Arctic*. New York: Macmillan Co., 1940.

THE RING AROUND THE MOON
MWR 42:446, 1914 "Haloes and Their Relationship to the Weather" A. H. Palmer
MWR 44:67, 1916 "Haloes at Fort Worth Texas and . . ." H. H. Martin
MWR 46:119, 1918 "Further Study of Haloes in Relationship to the Weather" H. H. Martin
SA 238: April/144, 1978 "Atmospheric Haloes" D. K. Lynch

THE CROSS IN THE SKY
Arctos 3:5, 1962 "The Christian Signs on the Coins of Constantine" P. Braun
MacMullen, Ramsay. *Constantine*. New York: Dial Press, 1969.

THE TURN OF THE TIDE
AMSB 2: Oct/14, 1921 "The Winds at the Turn of the Tide" H. M. Plummer
AMSB 3:50, 1922 "Rain at the Hour of High Water" F. M. Cutter
AMSB 25:369, 1944 "Is the Potomac a Barrier to Thunderstorms?" R. L. Feldman
Marine Observer 3:59, 1926 "The Tendency of the Wind to Change with the Tide"
MWR 32:59, 1926 "Deflection of Thunderstorms with the Tides" L. T. Garretson
MWR 93:93, 1965 "Diurnal and Semidiurnal Atmospheric Tides in Relationship to Precipitation Variations" G. W. Brier

THE MOON AND THE WEATHER
La Meteorologie 61: Sept/297, 1938 "The Reality of the Moon's Influence on the Weather" L. Rodes
La Meteorologie 71:229, 1928 "The Phases of the Moon and Weather Changes" M. Madrelle
La Meteorologie #68:341, 1962 "The Influence of the Moon on Changes on the Weather" C. Morel
New York Academy of Sciences: Transactions 27:676, 1964 "Lunar Tides, Precipitations, Variations and Rainfall Calendricities" G. W. Brier
Science 137:748, 749, 1962 "Lunar Synodical Period and Widespread Precipitation" D.A. Bradley, M. A. Woodbury, G. W. Brier

FARADAY'S WEATHER UPDATED
Journal of Meteorology 17:371, 1961 "An Electromagnetic Basis for the Initiation of a Tornado" I. R. Rathbun
Water Works and Sewage 82:186, 1935 "Electron Shells, Electron Atmospheres and Weather" H. P. Gillette

SOLAR WEATHER
Economist 280: July 18/85, 1981 "The Fickle Sun"
Shell Aviation News #214: April/19, 1956 "Utilizing the Sun to Forecast the Weather" E. D. Farthing
Science 250:481, 1972 "27 Day Cycle in the Rainfall at Los Angeles: R. L. Rosenberg, P. J. Colman

NORTHERN LIGHTS: SOUTHERN WINDS
Edinbourgh Philosophical Journal 12:83, 1825 "On the Aurora Borealis and Polar Fog" C. Hansteen
Magazine of Natural History 8:343, 1935 and 9:31, 1837 "Notices of 184 Observations on the Aurora Borealis . . ." W. H. White
Philosophical Transactions of the Royal Society 64:128, 1774 "Remarks on the Aurora Borealis" J. S. Winn
SA 33:182, 1894 "Sun Clouds and Rain Clouds" W. D. Roberts
SMM 33:182, 1894 "Aurorae and Storms" J. Slatter

EARTHQUAKE WEATHER

Cambridge Philosophical: Proceedings 9:351, 1895 "Electrical Properities of Newly Prepared Gasses" J. S. Townsend

Classical Weekly 23:2, 11, 1929 "Clouds, Rainbows, Weather Galls, Comets and Earthquakes as Weather Prophets in Greek and Latin Writers" W. S. McCartney

Report on the British Association 1, 1850 and 272, 1851 and 118, 1853 and 1, 1854 "Report on the Facts of Earthquake Phenomenon" R. Mallet

Societe Belge D'Astronomie Bulletin 13: Jan/44, 1980 "Earthquakes and Meteorological Phenomenon" E. L.

Tributsch, Helmut. *When the Snakes Awake*. Cambridge, Mass.: M.I.T. Press, 1982.

THE ELECTRICAL SKY

Physical Review 78:254, 1950 "Electrical Phenomena Occuring During the Freezing of Dilute Solutions . . ." E. J. Workman, S. E. Reynolds

SA 188: April/32, 1953 "The Earth's Electricity" J. E. McDonald

THE NEON SKY

Journal of the Scottish Meteorological Society 8:191, 1889 "St. Elmo's Fire at Ben Nevis" A. Rankin

Sea Frontiers 22:367, 1976 "The Fire of St. Elmo" K. C. Heidorn

Washington Academy of Sciences, Journal 45:269, 1955 "Pehr Kalm's Meteorological Observations in North America" E. L. Larson

RADIO WEATHER

AMSB 27:114, 1946 "Ionospheric Reflections and Weather Forecasting for Eastern China" E. Gherzi

Journal of Applied Physics 8:141, 1937 "The Daylight Variation of Signal Strength" R. C. Colwell, A. W. Friend

Physical Review 3:346, 1914 "Radio Transmission and Weather" A. H. Taylor

Proceedings of the West Virginia Academie of Science 3:234, 1929 "The Effect of Weather Conditions on Radio Reception" R. C. Colwell, R. G. Owens

THERMALS AND THUNDERSTORMS

MWR 42:348, 1914 "The Thunderstorm and It's Phenomena" W. J. Humphries

MWR 50:281, 1922 "The Local or Heat Thunderstorm" C. Brooks

Weatherwise 2:61, 1949 "Thunderstorm Today? Try a Probability Forecast" S. Price

THE MIGHTY HURRICANE

L'Astronomie 77:405, 1963 "The Dynamics and Structure of Solar Flares and Their Action on the Troposphere of the Planet" M. A. Djakov

Nature 165:38, 1950 "The Ionosphere and the Weather" E. Gherzi

SA 191: June/32, 1954 "Hurricanes" R. H. Simpson

Piddington, Henry. *The Sailor's Horn Book*. London: Williams and Norgate. 1860.

STRANGE RAINS

Bombay Natural History Society, Journal 69:202, 1972 "Rain of Fish in Shillong, Meghalaya" R. S. Pillai, S. J. S. Hattar

MWR 45:217, 1917 "Showers of Organic Matter" W. L. McAtee

SMM 21:144, 1887 "Remarkable Showers" H. S. Wallis

SMM 41:182, 1909 "Raining Cats and Dogs" B. T. Rouswell

LIGHTNING AND THUNDER

Classical Weekly 25:183, 200, 212, 1932 "Classical Weather Lore of Thunder and Lightning" E. S. McCartney

MWR 56:219, 1928 "Phenomena Proceeding Lightning" A. McAdie

MWR 59:481, 1931 "White Lightning Verses Red as a Fire Hazzard" W. J. Humphries

SMM 34:271, 1908 "Observations on the Color of Lightning Made at Epsom" S. C. Russell

GREEN LIGHTNING RODS

American Forest and Forest Life 32:259, 1926 "Trees Favored and Hated of Jove" E. W. Woolard

Botanical Society of Edinborough 6:71, 277 and 10:262 and 12:181, 497, 1873 'Notes on Lightning Striking Trees'
C. R. Academie de Sciences 175:1087, 1922 "Lightning and Trees" M. V. Schaffer
MWR 52:492, 1924 "Why an Oak is Often Struck by Lightning: A Method of Protecting Trees Against Lightning" R. N. Covert
SMM 1:74, 1896 "Lightning and Trees"

WIND, WATER AND WEATHER
Field and Stream 75: July/40, 1970 "Water Temperature is a Fishy Business" D. Richey
Field and Stream 82: July/58, 1977 "The pHish pHinder?" J. Scott
Fishing World 25: #2/30, 1978 "Penetrating the Mystery of the Thermocline" P. Moss
Journal of Meteorology 4:45, 1927 "Tracking Storms by Forrunners of Swell" W. H. Munk

THE SAILOR'S WEATHER
Classical Weekly 27:1, 9, 17, 25, 1933 "Greek and Roman Weather Lore of the Sea" E. S. McCartney
Mariner's Mirror 36:81, 1950 "By Weather to Vineland" C. A. Burland
Mariner's Mirror 56:219, 1970 "Loadstone and Sunstone in Medieval Iceland" B. E. Gelsinger
Josselyn, John. *Two Voyages to New England*. Boston: W. Veazie, 1865.
Poste, E. *The Skies and Weather: Forecasts of Aratus*. London: Macmillan & Co., 1880.

THE WEATHER FISH
Academie des Sciences; Memoires et Historie 424, 1666–86 "Observations on a Fish Resembling a Truite, Which is a Living Barometer" G. Clauder
Chemical and Engineering News 54: Sept 13/56, 1976 "Bermudians Use Shark Oil to Forecast Hurricanes"
Natur 48:426, 1899 "Weather Prophets in the Fishbowel" M. Danler
Proceedings of the Imperial Academy 8:375, 1932 "The Responces of the Catfish "Parasilures Asotus" to Earthquakes" S. Hatai, N. Abe

FISHING WEATHER
Attakapas Gazette 5: Sept/34, 1970 "Ask the Fish . . ." A. Simmons
Bombay Natural History Society Journal 48:598, 1948 "A Possible Cause of Blank Days When Mahseer Fishing" E. P. Gee
Field and Stream 34: June/32, 1929 "The Weather and Our Fishing" R. Bergman
Field and Stream 44: June/22, 1939 "Fish by Barometer" B. C. Snider
Kansas Academie of Science, Transactions 45:358, 1942 "Relationship of Barometric Pressure to Fishing Conditions" E. W. Huffman
Sky and Telescope 4: Aug/10, 1945 "Fish and the Aurora"

HUNTING WEATHER
Field and Stream 27: Oct/36, 1922 "Hunting With a Barometer" T. Bissell
Field and Stream 68: June/40, 1963 "Count in the Weatherman" D. L. Allen Outdoor Life 114: Sept/52, 1954 "Work With the Weather to Sound Out Your Deer" F. E. Sell
Outdoor Life 168: Oct/114, 1981 "The Statistical Squirrels" L. Mueller

INSECTS AND WEATHER
Archives des Sciences Physiques et Naturelles 44:413, 1917 "Influence of Atmospheric Pressure on the Development of Butterflies" A. Pictet
Bee Research Association 57: #2/50, 1976 "Effects of Electrical Charges on Honey Bees" U. Warnke
Mechanical Engineering 101: Sept/55, 1979 "Electrified Bees" E. Erickson
SMM 9:322, 1892 "Can Spiders Prognosticate the Weather" H. Mccook
SMM 21:27, 1886 "Ants as Rain Predictors"

THE PREDICTIVE LEECH
Annals of Philosophy 8:450, 1860 "On the Horse Leech as a Prognosticator of the Weather" J. Stockton
China Reconstructs 9: Jan/13, 1960 "Home-made Weather Forecasts" C. T. Li, C. Hu

Weather 22:288, 1967 "Observations on the Leech Worm . . ."
Both leeches and loaches (weather fish) may be obtained from Carolina Biological Company/
    2700 York Road/Burlington, North Carolina 27216

BIRD PREDICTIONS
American Speech 26:268, 1951 "Bird Names Connected with Weather, Seasons and Hours
    W. L. McAtee
MWR 26:354, 1898 "The Effect of Approaching Storms Upon Song Birds" C. E. Linney
SMM 12:9, 1877 "Influences of the Weather on Birds" W. C. Ley

ANIMAL FORECASTS
AMSB 11:32, 1930 "Observations Regarding the Effects of Air Pressure Upon Animal Life"
    H. P. Lasker
Classical Weekly 14:87, 97, 1920 "An Animals Weather Bureau" E. S. McCartney
Eclectic Medical Journal 12:436, 1853 "On the Indications of Weather as Shown by Animals,
    Insects and Plants" W. H. B. Thomas
MWR 48:98, 1920 "Animal Weather Prophets"

PLANTS AND WEATHER
Biological Bulletin 113:112, 1957 "Lag-lead Correlations of Barometric Pressure and Biological
    Activity" F. A. Brown, H. M. Webb, E. J. Macy
Clasiscal Weekly 17:105, 1924 "The Plant Almanac and Weather Bureau" E. S. McCartney
Fieldiana Botany 21:200, 1940 "Travels of Ruiz, Pavon and Dombrey in Peru and Chile"
QJRMS 55:15, 1926 "Notes on the Behavior of Certain Plants in Relation to the Weather" N. L.
    Silvester
Scientia 103:245, 1968 "Endogenous 'Biorhythmicity' Reviewed with New Evidence" F. A.
    Brown

THE WEATHER PLANT
Kew Bulletin 1, 1890 "Weather Plant, Abrus Precatorious" T. Dyer
Scientific American Supplement 65:280, 1908 "The First Weather Plant Observatory: A Novel
    English Experiment" H. J. Shepstone
Callahan, Phillip. *Tuning Into Nature.* Old Greenwich, Conn.: Devin-Adair Co., 1975.

THE OAK AND THE ASH
Folklore 66:296, 1955 "Oak Before Ash" W. B. Whitworth
SMM 15:9, 1880 "The Oak and the Ash" E. Shapman
SMM 24:97, 120, 121, 1889 "The Oak and the Ash" T. A. Chapman, F. E. Evans, E. Lees

PRACTICAL PHENOLOGY
Agroborealis 12: Jan/42, 1980 "Using Phenology to Characterize Spring Seasons in Alaska"
    W. M. Mitchell
Journal of Agriculture Research 20:151, 1920 "The Influence of Cold in Stimulating the
    Growth of Plants" F. V. Coveille
MWR Supplement #9, 1918 "Periodic Events and Natural Law as Guides to Agriculture
    Research and Practice" A. D. Hopkins
Outdoor Life 159: April/90, 1977 "Nature is My Calendar" B. W. Dalrymple

RAIN TREES: RAIN FORESTS
Bulletin of the Torrey Botanical Club 3:38, 1872 "Trees and Rain" J. Merrian
Bulletin of the Torrey Botanical Club 4:5, 1873 "Trees and Rain" F. Hubbard
Revue Tunisienne 33:67, 1927 "The Forests of the Sahara" L. Lavauden
Science 197:7, 1977 "The Amazon Basin, Another Sahel?" I. Friedman
Unasylvia 17: #11/13, 1963 "Forest History of the Near East" K. H. Oedekoven
Unasylvia 21: #84/28, 1967 "Acacia Peuce; A Tree for Arid Areas" L. D. Pryor

HUMAN WEATHER REACTIONS
American Journal of Physical Medicine 37:83, 1958 "Biologic Effect of Ionized Air in Man"
    T. Winsor, J. C. Beckett

Archives of Environmental Health 12:279, 1966 Suicides and Climatology" E. Digon, H. B. Bock

Journal of Gerontology 14:344, 1959 "Air Ionization and Maze Learning in Rats" J. Jordan, B. Sokoloff

Journal of Personality and Social Behavior 39:1947, 1979 "Quasi Experiments with the Sunshine Samaritan" M. E. Cunnington

WEATHER DREAMS

Journal of Psychiatric Nursing and Mental Health Services 14:13, 1976 ". . . Confirmation and Exploration of the Transylvania Effect" S. H. Geller, H. Shanon

Science 143:263, 1963 "Sleep Tendencies: Effects of Barometric Pressure" W. B. Webb, H. Ades

Chetwynd, Tom. *A Dictionary for Dreams.* London: Allen and Unwin. 1972.

Anderson, George K. *The Sage of the Volungs.* Newark, Delaware: University of Delaware Press, 1982.

ARTHRITIS PREDICTIONS

American Journal of Medical Science 73:305, 1877 "Relationship of Pain to the Weather . . ." S. W. Mitchell

International Journal of Biometeorology 10:105, 1966 "Possible Relationship Between Weather Hexosamine Excretion and Arthritis Pain" S. W. Troomp, J. Bouma

Journal of the American Medical Association 92:1995, 1929 "Arthritic Pain in Relation to Changes in the Weather" E. Rentschler, F. Vanzart, L. Rowntree

Weather 10:183, 1955 "Bunions and Seaweed" A. J. Whiten

MYSTERY OF WEATHER BEHAVIOR

Biometeorology 6:pt. 1/126, 1957 "Significant Correlations Between Human Anxiety Scores and Prenatal Activity" M. A. Persinger

International Journal of Biometeorology 12:263, 1968 "The Psychological Consequences of Exposure of High Density Pulsed Electromagnetic Fields" F. G. Hirsch, D. R. McGiboney, T. D. Harnish

Meteorological Abstracts 16:2–98, 1964 "Experimental Study of the Effect of Weather on a Group of Fourth and Fifth Grade Pupils" C. B. Gedeist

THE WINTER WEATHER GAME

American Journal of Arts and Science 2:255, 1820 "May Not the State of Plants which Bloom Late Indicate a Late Autumn" C. Deway

American Naturalist 48:122, 1914 "Humidity—A Neglected Factor in Environmental Work". F. Lutz

Bulletin de Scientifique Pharmacie 36: Supp/65, 1929 "The Bees Predict a Hard Winter" A. Guillaime

Classical Weekly 16:3, 1922 "The Folk Calendar of Times and Seasons" E. S. McCartney

Cycles 18:23, 1967 "The Bears" S. Stevens

MWR 57:455, 1929 "The Influence of Weather Factors in India on the Following Winter in Canada" F. Groissmayr

THE SPRING WEATHER GAME

SA 136:269, 1927 "The Ground Hog is a Poor Prophet"

Weather 17:13, 1962 "Candlemas Day" M. S. Roulston

Weather 21:433, 1966 "Weather Lore for the Twelve Months" E. R. Yarham

PRAYING FOR RAIN

Christian Century 47:1084, 1930 "Does Prayer Change the Weather?"

Masterkey 29:94, 1955 "Notes on Yokuts Weather Shammanism and the Rattlesnake Ceremony" F. A. Riddell

San Antonio Light Aug/22:9C, 1948 "Was It Snakes or Prayer that Brought Rain" J. F. Dobie

San Antonio Light Feb/23:7B, 1964 "The Drought and Prayer" J. F. Dobie

ANCIENT RAINMAKING

Australian Musuem Magazine 10:249, 302, 1951 "Aboriginal Rain Makers and Their Ways" F. W. McCarthy

Fate 4: Nov/24, 1951 "Rain from the Hopi Snake Dance" O. E. Singer
Fate 10: April/57, 1957 "Could the Sioux Control Weather" F. J. Goshe
National Geographic Magazine 6:35, 1894 "Weather Making Ancient and Modern" M. W. Harrington
DeRiencourt, Amaury. *Roof of the World.* New York: Rinehart, 1950.

MODERN RAINMAKING
Journal of Orgonomy 19: May/57, 1985 "Field Experiments with the Reich Cloudbuster" J. De Meo
Popular Science 129: July/34, 1936 "Weird Scheems to Make It Rain" R. E. Martin
Science 180:1347, 1973 "Weather Modification: Colorado Heeds Voters in Valley Dispute" J. L. Carter
Kelley, Charles. *A New Method of Weather Control.* Stamford, Conn.: published by author, 1961.

THE GREAT FLOODS
Annual Report of the Smithsonian Institution 325, 1938 "The Meteorology of Great Floods in the Eastern United States" C. F. Brooks, A. H. Thiessen
Journal of Geology 97:504, 1969 "The Lake Missoula Floods and the Channeled Scabland" H. Bretz
MWR 88:25, 1960 "Mean Monthly Values of Precipitable Water" C. H. Reutan
Popular Science 102: Feb/44, 1923 "When the Bottom Fell Out of the Sky" J. E. Hogg

WEATHER WORDS
American Speech 5:222, 1929 "Blizzard Again" A. Read
Weather 1:146, 1945 "A Plea for the Abolition of 'Meteorologist' " S. Chapman
Weather 26:536, 1971 "The Meteorology of Anglo-Saxon Month Names" E. N. Lawrence
"The Oxford English Dictionary"

INDIAN SUMMER
MWR 30:19, 69, 1902 "The Term Indian Summer" A. Matthews
MWR 39:469, 1911 "Indian Summer" J. Morrow
MWR 44:575, 1916 "Indian Summer and Plimsoll's Mark" W. R.
Proceedings of the American Philosophical Society 62:48, 1923 "The Indian Summer as a Characteristic Weather Type . . ." R. De Ward

MAKING MONEY ON THE WEATHER
Cycles 13:123, 1962 "A Significant Experiment on Temperature Range and Rainfall: E. L. Smith
Cycles 17:66, 1966 "Effect of Sunspot Activity on the Stock Market" C. J. Collins
Fortune 95: April/59, 1977 "The Weather and the Future's Market" L. Snyder
Journal of the Meteorological Society of Japan 29:130, 1951 "Influence of Weather on Mental States and the Stock Market" S. Yamaguchi
Sales Management 82: April 3/93, 1959 "Subtle Weather Changes"

THE WEATHER KITE
Annual Report of the Smithsonian Institution 223, 1900 "The Use of Kites to Obtain Meteorological Observations" A. Rotch
MWR 21:418, 1895 "A Weather Bureau Kite" C. F. Marvin
MWR 50:163, 1922 "An Aerological Survey of the United States" W. R. Gregg
Weather 36:294, 1981 "Kites and Meteorology" G. J. Jenkins

INDEX

Abrus precatorius, weather sensitivity of, 155–156

Absolute humidity, 27

Acosta, Joseph, 8

Acroclinium roseum, weather sensitivity of, 151–153

Adams, John Quincy, 100

Aerodynamics, 205

Air masses
concept of, 42–43
theory of, 33

Air thermometer, first, 24

Anaximander, 12

Andronikos, 15

Animal forecasts, 146–148

Animals
impact of weather on behavior of, 173
and prediction of weather, 45
weather sensitivity of, 132–135, 146–148, 180

Ararat, Mount, 6

Aratus, 117

Archibald, Douglas, 213

Aristophanes, 189

Aristotle, 12, 16, 19, 47, 84, 175

Arthritis, forecasts of, 178–180

Arthritis weather, 178

Asamusi Marine Biological Station, 127–128

Ash tree, weather sensitivity of, 159–161

Athos, Mount, rain cap of, 52

Aurelius, Marcus, 190

Aurora borealis *See* Northern lights

Aztecs, and wind measurement, 42

Backing winds, 42

Baden-Powell, Robert, 212

Bad weather signs, 120

"Ballad of Sir Patrick Spence", 60

Baquiros, 102

Barometer(s)
invention of, 19
natural, 21–23
secrets of the, 19–20

Barrow, Sam, 192

Bartles, Robert, 59

Bartold, Sebastiano, 25

Bathtub, weather, in, 37–39

Bears, weather sensitivity of, 148, 183

Bees, weather sensitivity of, 138

Beetles, weather sensitivity of, 137

Bellow, Saul, 197

Bengal cyclones, 102

Big Ben, as weather bell, 44

Bioclimatic Index, 211

Birds
migratory behavior of, 145
weather sensitivity of, 143–145

Bishop's Ring, 54

Bjerkness, Jacob, 42–43, 74
air mass theory of, 100–101

Bjerkness, Wilhelm, 42–43, 74
air mass theory of, 100–101

Black Hills, 21

Blizzards, 120–123, 205

Bluebirds, weather sensitivity of, 145

Blue horizon, as sky phenomena, 54

Boori, and prediction of wind change, 127

Borland, Hal, 148

Bose, Jagadis, 158

Bouffier, Elzeard, 169

Boyle, Robert, 66

Bradford, William, 69

Bretz, Harlen, 202

Brown, Dr. Frank, 154–155

Brown, Will, 137

Buckmoth, 165

Bunion Brigade, 180

Burroughs, John, 183

Butterflies, weather sensitivity of, 184

Buys-Ballot, Christopher, 31, 39

Byrd, Richard, 58

Caesar, Julius, 15, 162

Calendar migrants, 145

California fogs, production of, 39

Callighan, Philip, 158

Cartesian diver, 22

Castor and Pollux, 91–92

Caterpillars, weather sensitivity of, 136–137, 139, 183–184

Catfish, prediction of earthquakes by, 127–128

Cats, weather sensitivity of, 148

Chantry, Sir Francis, 130

Charlemagne, 40

Chiloe Islands, 27

Cicadas, weather sensitivity of, 138

Cirrus clouds, 52

Cleopatra's Needle, 28

Climate, effect of volcanoes on, 3–4

Clockwise winds, 42

Clouds
  predictions in, 50–52
  types of, 50, 52

Cloud-seeding research, 199

Coal mines, as singing mines, 21

Cobalt chloride, as rain predictor, 27

Cockroaches, weather sensitivity of, 139

Cold fronts, 101

Cold waves, 122

Coleman, Paul, 78

Collinson, Peter, 212

Columbus, Christopher, 4–5, 13, 60, 92, 102

Compass, 40, 118

Constantine, 63, 65

Coriolis, George, 37, 39

Coriolus force, 37–39, 75, 76

Cottonwood tree, weather sensitivity of, 154

Counterclockwise winds, 42

Cowper, William, 140

Cricket temperature, 25

Cross in the Sky, 63–66

Cumulus clouds, 52

Curran, Charles, 184

Curtius, Quintus, 168

Cyclone, 205

Dalton, John, 81, 82

Dandelion, weather sensitivity of, 151

Darwin, Charles, 31, 33, 155

Da Vinci, Leonardo, 3, 8

Dee, John, 30

Deer, weather instincts of, 132–33, 35

Democritus, 12, 210

De Oviedo, Fernando Huracan, 102

DeSaussure, William, 28

Diocletian, 63

Diodorus, 16

Djakov, Anatoli, 104

Dobie J. Frank, 192

Dolbear, A. E., 25

Dolbear's Law, 25

Donner, George, 123

DOR (deadly orgone radiation), 198

Dreams, weather symbolism in, 175–77

Ducks, weather instincts of, 135

Du Fay, Charles, 20

Dunn, Elias, 121

Dust Bowl, 169

Dutch Weather House, 30

Earth, rotation of, and weather, 37–39

Earthquakes, 76, 84–87

Eddies, 74

Edwards, Robert, 211

Electrical sky, 87–90

Elk, 165

El Ninō, impact of, on rainfall, 202

Empedocles, 12

Epilion layer, 115–116

Espy, William, 99–100, 213

Euonymus radicans coloratus, as thermometer, 25

Eye, of a hurricane, 103

Ezekiel's wheel, 62

Fabre, Jean, 137

Fairweather Glacier, 60

Faraday, Michael, weather theories of, 74–76

Ferdinand II, 20

Fishing, water temperature in predicting fish location, 116

Fitzroy, Robert, 31–33

Flies, weather sensitivity of, 138

Floki, 118

Floods, 6–10, 200–203

Florentine thermometers, 24

Fog horn, 44

Forecast, early, 30–33

France, Anatole, 173

Franklin, Benjamin, 3–4, 82, 87, 212–213

Freeman, James, 207

Freud, Sigmund, 197

Frost fish, 127

Fuji, Mount, rain cap of, 52

Fulgurites, 110

Full moon, impact of, on human behavior, 176

Furtenbach, Joseph, 37

Galen, 23
Galileo, 19, 24
Galton, Francis, 190
Geosmin, 49
Gherzi, Ernest, 96, 104
Gilgamesh, 7
Giono, Jean, 169
Govinda, Lama Anagarika, 176
Grant, Joan, 193
Green flash, as sky phenomena, 54
Green lightning rods, 111–114
Green sky, as indication of snow, 54
Grey, James, 192
Ground hog, myth of, 186–189
Ground Hog Club of Punxsutawney, Pennsylvania, 187

Hair, as natural hydrometer, 28, 30
Halcyon days, 205
Half winter, 208
Hargrave, Lawrence, 213
Hargrave kite, 213
Harmar, General, 208
Hatfield, Charles, 196–197, 197
Hawks, weather sensitivity of, 144–145
Heat thunderstorms, 101
Hecataeus, 12
Heckwelder, John, 163
Heiland, Fritz, 64
Helena, 91–92
Herschel, Sir William, 83
Hess, Victor, 88
Hessian fly, 164
Hexosamine, 179
Hillengar Effect, 57, 60
Hollow earth, legends of, 58–59
Homer, 40
Hooke, Robert, 20, 31, 42
Hopi Indians, and rain making, 195
Horologion of Andronikos, 14–16
Hot weather, impact of on human behavior, 173–174
Howard, Luke, 50, 72
Human behavior, impact of weather on, 173–175
Human weather reactions, 173–175
Humboldt, Alexander von, 84–85, 107, 154
Humidity, 179
Hunrakan, 102
Hunting, determining best weather for, 132–135
Huroc tribe, and rain making, 195

Hurricane birds, 104
Hurricane flower, 104
Hurricanes, 102–105
    prediction of, by Seminole Indians, 5–6
    use of shark-oil for predicting, 129
Hydormeters, natural, 27–30
Hydrodynamics, 205
Hydrometry, measuring water, 27
Hypolimnion layer, 115–116

Ice blink, 58–59
Indian summer, 206–209
Insects, weather sensitivity of, 136–139
Inversion layer, 26
Ionosphere, 94
Ions
    effect of, on sinus conditions, 174
    impact of, on human behavior, 182

Jacob's ladder, 62
Jalbert, Domina, 215
Japan, blooming of cherry trees in, 165
Jefferson, Thomas, 3
Jevons, Stanley, 210
Jones, Peter, 207
Josephus, 62
Josselyn, John, 118

Kalm, Peter, 91
Kant, Immanuel, 85
Kelly, Charles, 198
Kenyatta, Jomo, 166
Killdeer Mountains, 21
Kites, weather sensitivity of, 145
Knight, John Alden, 131–132, 134–135
Krefting, Otto, 180
Kreuger, Paul, 13
Kyritsis, Mathon, predictions, 185

Lactantius, 13
Lakes, role of winds and air temperatures in freshwater, 115–116
Lakhovsky, George, 181
Lama, Dalai, 195
Lamarck, Jean Baptiste, 50
Lapse rate, 25, 212
Leahy, Frank, 193
Leech, weather sensitivity of, 140–142
Leopold, Prince, 24
Lightning, 108–110
Lightning rods, trees as, 111–114
Lindbergh, Charles, 104

Lithodes antarcticus, shell of, as natural hydrometer, 27
Little Ice Age, 78
Lloyds of London, 31
Loach, 128
Lomonosov, Michael, 19
Lucian, 44
Lucretius, 108
Lunar declination cycle, 72–73
Lunar moon, 68
Lutz, Dr. Frank, 184

Magellan, Fernando, 92
Magnesium chloride, as rain predictor, 27
Malay kite, 213
"Mappa Mundi", 12
Matsukaze, 44
McAdams, John, 146
McCook, Henry, 137
Medicine Hole, 21
Melatonin, 173
Melbourne, Frank, 196
Merl, William, 30
Merriam, James, 166
Merryweather, George, 142
Meteorites, 62
Meteorology, 204
Meterologists, adoption of wind rose by, 41
Mierback, Karl, 104
Mines, as natural barometers, 21–22
Mirage phenomenon
   Hillingar Effect, 57, 60
   and thermal stratifications, 54, 56
Mitchell, Weir, 178
Moon
   ring around, 60–62
   and the weather, 70–73
Moor, Henry, 210
Morran, Charles, 162
Mosquitoes, weather sensitivity of, 139
Mountains, rising of the, 57–60

Naddod, 118
Nash, Richard, 197
Navarra, Fernand, 10
Neon sky, 91–93
Nevis, Ben, 92
Newton, Isaac, 3, 42, 66
Noah's Flood, myth of, 6–10, 202
Northern lights, 62, 76, 80–83
Northers, 122
Nowack, Josef, 156, 158

Oak tree, weather sensitivity of, 159–161
Ocean currents, 39
Osteoarthritis, 179–180
Ovando, Governor, 4
Ox-eye daisy, weather sensitivity of, 151

Packer, Alfred, 123
Parhelion, 63
Pascal, Blaise, 19
Paulinus, Suetonius, 168–169
Perier, Florin, 19
Perry, Matthew, 58
Petrichor, 47, 48
Phenology, 159, 161
   practical uses of, 162–165
Piddington, Captain, 205
Plants, weather sensitivity of, 151–169
Plato, 168, 175
Plimsoll, Samuel, 209
Pliny the Elder, 27–28, 47, 84, 105, 137, 146
Polo, Marco, 194
Poplar tree, weather sensitivity of, 154
Porlieria lorentzii, 153
Potato, weather sensitivity of, 154–155
Prairie dogs, weather sensitivity of, 185–186
Probability forecast, 101

Quartpot, Robin, 195
Quatremere-Disjonval, Denis, 137–138

Rabbits, weather instincts of, 134
Radio signal weather forecasting, 96
Radio status, effects of, on health, 181–182
Radio weather, 93–95
Rain
   praying for, 189–193
   smell of, 47–49
   strange, 105–107
Rain-caps, 52
Rain dances, 194–195
Rain forests, 166–169
Rainmaker, The, 197
Rainmaking
   ancient ceremonies for, 194–195
   modern attempts at, 196–199
   scientific, 198–199
Raleigh, Sir Walter, 9
Rattlesnake weather lore, 148
Redfield, William, 39, 122
Reich, Wilhelm, 197–199

Relative humidity
  definition of, 25
  measuring, 27
Rheumatoid-arthritis, 180
Rhododendron leaves, 25
Robins, weather sensitivity of, 145
Rosenburg, Ronald, 78
Ross, William, 69
Royal Society of London, 30
Ruiz, Hipolito, 153

Saga of Guomund, 176
Sahara desert, creation of, 169
Sailors, predictions of weather by, 117–120
St. Andrew's Cross, 64
St. Augustine, 13
St. Elmo's fire, 91–92
St. Eracemus, 92
St. John, Bishop, 176
Sales Productivity Services, 211
San Francisco earthquake (1906), 86
Santa Ana winds, 174
Scarlet pimpernel, 151
Scientific Society of Rochdale, England, 25
Scum
  formation of, 22
  as sign of barometric pressure, 22
Sea gulls, weather sensitivity of, 145
Season, Henry, 140
Seasonal dimorphism, 184
Seaweed, as rain predictor, 27
Seminole Indians, prediction of hurricanes by, 5–6
Sferics, 93, 94
Shad bush, 163
Shakespeare, 173
Shark-oil, use of, for hurricane prediction, 129
Shark-oil barometers, 129
Sin, Han, 212
Sinclair, George, 20
Sky phenomena, as weather predictor, 53–54
Slocum, Joshua, 13
Smith, Adam, 210
Smith, Edgar, 211
Smog, 205
Smoke, as weather predictor, 25–26
Snakes, weather sensitivity of, 148
Snow fires, 91
Solar weather, 77–80
Solunar theory, 131–132, 134

Soyal festival, 195
Spence, Patrick, 60
Spiders, weather sensitivity of, 137
Spring, prediction of arrival of, 186–189
Squaw winter, 208
Squirrels, weather sensitivity of, 134, 185
Stock market, use of weather in predicting behavior of, 210–211
Stockton, James, 141–142
Storm(s), 205
  barometric prediction of, 20
  swirl of winds in, 39
Storm lights, 91
Storm warnings
  blizzards, 120–123
  hurricanes, 102–103
  lightning, 108–110
  rain, 105–107
  sailor's, 117–120
  thermals, 99–101
  thunder, 108–110
  tides, 114–116
  trees as lightning rods, 111–114
  waves, 114–116
Sun pillar, 65
Sunset
  concept of red, 53, 54
  concept of yellow, 53, 54
Sun spots, 77–78
  cycle of, 72–73
Sunstone, 118
Swain, James, 213
Swifts, weather sensitivity of, 144
Sylvester II, 13

Termites, weather sensitivity of, 138
Terreros, Captain, 4
Theophrastus, 12
Thermals, and thunderstorms, 99–101
Thermocline layer, 115–116
Thermometers, 23–26
Thomas, Guboo Ted, 138
Thomas Jefferson's Garden Book, 3
Thoreau, Henry David, 4, 208
Thunder, 108–110
Thunderstorms, 200
  and thermals, 99–101
Tidal phenomena, 66–70
Toadfish, 127
Tornado, 76–77, 148
Torricelli, 19
Tower of the winds, 14–16

Traditional lore, 188
Transylvania effect, 176
Trees
  electrical effects of, 114
  hit by lightning, 113
  as lightning rods, 111–114
Troldjol, Mount, 180
Tropical storms, 103
Turmlirz, Ottokar, 37
Typhoons, 102, 205

United States Navy, weather forecasts by,
    33
United States Weather Bureau, 33
  kite-flying program of, 213
Ussher, Archbishop, 7

Varro, 15
Veering winds, 42
Vikings, 118
Virgil, 81
Vitruvius, 15
Volcanic activity, 3–4, 85
Von Guericke, Otto, 20
Vortice shedding, 44

Wadsworth Longfellow, Henry, 207
Warner, William, 143
Washington, George, 3
Water bells, 43
Waves, prediction of storms by, 114–116
Waxwings, weather sensitivity of, 183
Weather
  anecdotes on, 3–6
  ancient, 11–13
  in the bathtub, 37–39
  history of the study of, 3–16
  sounds, 43–46
  terminology, 204–206
Weather behavior, 180–182
Weather bell, Big Ben as, 44
Weather cycles, influence of noon on, 72
Weather fish, 127–129
Weatherford, C. P., 197
Weather kite, 212–215
Weather migrants, 145
Weather phenomena
  cloud predictions, 50–52
  cross in the sky, 63–66
  earthquakes, 84–87

electrical forces, 74–77
electrical sky, 87–90
mirages, 57–60
moon, 70–73
neon sky, 91–93
northern lights, 80–83
radio weather, 93–95
ring around the moon, 60–62
sky phenomena, 53–56
smell of rain, 47–49
solar phenomena, 77–80
tides, 66–70
water rotation, 37–39
weather sounds, 43–46
wind direction, 40–43
Weather plant, 155–158
Weather proverbs, oak and ash trees as
    subject of, 159, 161
Weather tools, 17–95
  barometers, 19–23
  Hydrometers, 27–30
  thermometers, 23–26
Wells, as natural barometers, 21
White, John, 127
Willheimer, John, 187
William the Conqueror, 117
Willy-willies, 102
Wilson, Alexander, 212
Wimsatt, Gordon, 148
Wind(s), 210
  direction of, 40–43
  swirl in, 39
Wind Cave, 21
Wind rose, adoption of, 40–41
Wind speed device, development of, 42
Winter
  need of plants for, 165
  predicting severity of, 183–186
Winter folklore, 186
Winter weather game, 183
Wire predictions, of weather, wires, 44–
    45
Wood, Fergus, 69
Woolley, Leonard, 7
Woolly bears, 183

Yellowstone National Park, geysers in, as
    natural barometers, 23
Yokut tribe, and rain making, 193